THE EVERYTHING ANSWER BOOK

MAR 2 0 2018

Ms. D. R. Toi
223 Southlake Pl.
Newport News, VA 23602-8323

The Everything Answer Book

HOW QUANTUM SCIENCE EXPLAINS
LOVE, DEATH
AND THE MEANING OF
LIFE

AMIT GOSWAMI, PHD

*I dedicate this book to the
quantum activists—past, present, and future—of the world.
We shall overcome.*

Copyright © 2017
by Amit Goswami
All rights reserved. No part of this publication may be reproduced or transmitted in any form or by any means, electronic or mechanical, including photocopying, recording, or by any information storage and retrieval system, without permission in writing from Red Wheel/Weiser, LLC. Reviewers may quote brief passages.

Cover design by Jim Warner
Cover photograph © R. T. Wohlstradter/shutterstock
Interior by Timm Bryson, em em design
Typeset in Garamond Pro

Hampton Roads Publishing Company, Inc.
Charlottesville, VA 22906
Distributed by Red Wheel/Weiser, LLC
www.redwheelweiser.com

Sign up for our newsletter and special offers by going to *www.redwheelweiser.com/newsletter/*.

ISBN: 978-1-57174-762-4
Library of Congress Cataloging-in-Publication Data available upon request

Printed in Canada
MAR
10 9 8 7 6 5 4 3 2 1

Contents

Acknowledgments vii
Introduction. ix

Chapter 1. A Clash of Two Worldviews 1
Chapter 2. Consciousness and the Science of Experience 15
Chapter 3. The Physics of the Subtle 37
Chapter 4. Zen and Quantum Physics 55
Chapter 5. Thought, Feeling, and Intuition 67
Chapter 6. The World of Archetypes 79
Chapter 7. The Ego and the Quantum Self 95
Chapter 8. Free Will and Creativity 101
Chapter 9. Involution and Evolution 119
Chapter 10. A Tale of Two Domains 129
Chapter 11. The Creative Principle 143
Chapter 12. Quantum Reincarnation 153
Chapter 13. The Meaning and Purpose of Life 173
Chapter 14. The Meaning of Dreams 187
Chapter 15. Enlightenment 195
Chapter 16. Quantum Professor, Quantum Society 207

Glossary. 229
Further Reading. 237

Acknowledgments

I thank Voice, Inc. for inviting me to Tokyo, and Masumi Hori for the dialogs I had with him. I thank Tatiana Hill for transcribing the tapes of those interviews. I thank Eva Herr for the interview that I had with her and many other journalists whose names I can't recall for their contributions as well. I thank Judith Greentree for a thorough reading of the manuscript and for some humorous comments that I incorporated into the book. Heartfelt thanks are due to Sara Sgarlat, Mimi Hill, and Terry Way for their contributions. I thank the editorial staff of Hampton Roads for a wonderful job of production. I thank you all.

Introduction

Almost one hundred years have passed since the complete mathematical formulation of quantum physics. It has been verified by myriad experiments and its concepts have been successfully applied in many technologies. Indeed, we have begun to use the word "quantum" in our daily discourse—often without fully understanding its deeper meaning. And yet, despite its effective integration into our society, the quantum worldview has still not been fully accepted by the scientific community, which continues to espouse and defend the archaic Newtonian worldview. Consequently, the full implications of the quantum worldview have not yet penetrated the public mind. The good news is that, in the 1990s, thanks to the efforts of an avant-garde group of renegade scientists including myself, the quantum worldview began to mature and gave birth to an all-inclusive new scientific paradigm. A grass roots movement known as "quantum activism" has begun to dislodge the stranglehold of Newtonian physics on the scientific establishment by appealing directly to people. This book is a part of that movement and the latest popular exposition of the quantum worldview.

Part of the mischief derives from circumstances. The prevailing Newtonian paradigm was always fraught with paradoxes. Officially known as scientific materialism, this worldview proposed that everything exists merely as a phenomenon of matter—material movement in space and time, caused by material interaction. The paradoxes implicit in this view were never resolved. It wasn't until the 1980s and 1990s that scientific materialism came under serious scrutiny by the scientific community, prompted by new experimenal data. Previously, the worldview of scientific materialism was much aided by the shift of physics away from a philosophy-oriented European approach to the more pragmatic American approach that followed World War II. Before the 1950s, scientific materialism was firmly entrenched only in the dicisplines of physics and chemistry—the science of inanimate objects. After the 1950s, it also began to dominate biology (which became chemistry), the health sciences (which became almost "mechanical"), and eventually psychology (which became cognitive neuroscience).

The second party to the mischief was the inadvertent enthusiasm of well-meaning scientists to close off the debate surrounding the meaning of quantum physics as quickly as possible. So, a compromise—dubbed famously (or should I say, infamously) the Copenhagen Interpretation—was reached. This interpretation was pioneered by the famous and amicable Niels Bohr, whom every physicist (including myself) worshipped.

The centerpiece of the Copenhagen Interpretation is called the "complementarity principle," which, in its popularized form, is simply wrong, both theoretically and experimentally. Quantum mathematics says unequivocally that quantum objects are waves. But of course, experiments say that they are also particles. How can the same object be both a wave—something that spreads out—and a particle—something that travels in a defined trajectory? The popular form of the complementarity principle resolves this wave-particle paradox by claiming that quantum objects are both waves *and* particles. The wave aspect is revealed in wave-measuring experiments; the particle aspect is revealed in particle-measuring experiments. But both aspects never show up in the same experiment and are thus called complementary.

However, the correct answer to the paradox of wave-particle duality—both theoretically and experimentally—is this: Quantum objects are waves of possibility residing in a domain of reality outside of space and time called the domain of potentiality. Whenever we measure these objects, they reveal themselves as particles in space and time. So both the wave and the particle aspects of an object *can*, in fact, be detected in a single experiment. Unfortunately, the popularized version of the complementarity principle, which created the impression that the wave and particle aspects of an object *both* exist in space and time, misled an entire generation or two of physicists and caused them to close their minds to the really radical elements of quantum physics. In fact, quantum physics insists on a two-level reality, not the single space-time reality of Newtonian physics and scientific materialism. Moreover, quantum physics cannot possibly be made paradox-free without explicitly invoking consciousness.

But of course, it was the role of consciousness that kept the paradox alive—not in the mainstream, but in a cultish sort of way. In the 1980s, an experiment by Alain Aspect and his collaborators resolved the issue of a dual versus a single domain of reality by discerning the domain of potentiality from the domain of space and time. In the former, no signal is needed for communication; everything is instantaneuously interconnected. By contrast, in space and time, signals, always moving with a speed no greater than the speed of light, mediate communication, which always occurs in finite time.

What does it mean to say that the domain of potentiality is all instantaneously interconnected? Simply this: Everything in the domain of potentiality is one entity. In a scientific paper published in 1989, and again in 1993 in *The Self-Aware Universe*, I arrived at the paradox-resolving proposition that the domain of potentiality is *our consciousness*—not in the form of ordinary ego-consciousness, but as a higher consciousness in which we are all one. In manifest awareness, we become separate partly due to the necessity of distinction from other objects (the subject-object distinction) and partly due to our individual conditioning. I also proposed that this higher One consciousness is causally empowered by *downward causation*—the capacity to choose among the many facets of a wave of possibility. It is conscious choice that transforms *waves* of possibility into *particles* of actuality.

Philosopher and scientist Willis Harman, at the time president of the Institute of Noetic Science (IONS), was very supportive of my work. He invited me to write a monograph on my research. The new research soon created a new science—"science within consciousness," a term I later discovered was already in vogue thanks to Harman. A monograph by the same name was published by IONS in 1994.

Progress in the field came rapidly and was always accompanied by strange coincidences of Jungian synchronicity. First, an old woman called me on a radio talk show with the question: What happens when we die? I didn't know how to answer her without resorting to cultural clichés, so I kept quiet. Then a Theosophist—a believer in reincarnation—took a course from me based on my book, *The Self-Aware Universe*, but ended up mostly talking about reincarnation. Soon after, I had a dream in which I woke up remembering this admonition: *The Tibetan Book of the Dead* is correct; it's your job to prove it. Finally, a graduate student of philosophy called me and asked me to help her mourn and overcome the impact of her boyfriend's death.

It was while conversing with her and trying to theorize about what survives us in death that I began to see the possibility of a science of *all* our experiences—material sensing (sensation), vital feeling (energy), mental thinking (meaning), and supramental intuitions (archetypes like love and truth). From this, I developed a theory of survival after death and reincarnation. Soon after, I got a call from author and editor Frank de Marco asking me to write a book on my newest research. This appeared in 2001 under the title *Physics of the Soul.* BK

Biophysicist Beverly Rubik called me in 1998 and asked me to contribute an article on my research to an anthology she was compiling. In 1999, I joined a group of thirty new-paradigm thinkers at a conference with the Dalai Lama in Dharamsala, India. This conference turned fractious. First, physicist Fred Alan Wolf and I had a verbal battle over whose approach to the new paradigm was correct. Others joined in; the organizers complained to the Dalai Lama. He simply laughed and said: "Scientists will be scientists." After peace was established, the Dalai Lama asked us to apply our new paradigm to social issues. This caught my attention. When I returned to the United States, I wrote the article Beverly Rubik had requested, applying quantum physics to health and healing. Here, I developed a theory of what Deepak Chopra called "quantum healing"— spontaneous healing without medical intervention.

Around the same time, I visited Brazil, where a young man asked me if I knew Deepak Chopra. When I said I did not, he said, "I can correct that." Soon after, I got an invitation to visit Deepak in San Diego. He had just published his book *Perfect Health* (2000), which discussed Ayurveda, an alternative healing system from India. He gave me a copy and asked me to read it.

As a result, I ended up proving the scientific validity of an idea that physicians of alternative medicine have been using for millennia. Since we are more than our physical bodies, diseases in our "subtle" bodies can also be responsible for physical disease, especially chronic disease. And thus healing can be approached, not only through curing physical symptoms, but also through healing disease at its more subtle source.

Practitioners of the health sciences, physical and mental, deal with actual human beings. Thus they do not always give their enthusiastic approval to the allopathic model of medicine—the more "mechanical" model that grew out of scientific materialsim. When I wrote *The Quantum Doctor* (2004), which dealt with integrating conventional "mechanical" medicine and more human alternative medicines, the quantum worldview began to get some traction among alternative medicine practitioners and even among some avant-garde allopaths. Deepak was so enthusiastic about the book that he wrote the foreword to a later edition.

Medicine is based on biology. To relax the stranglehold of scientific materialism on medicine, we must introduce consciousness into biology. I began that work in the 1990s and, in 2008, proposed a consciousness-based scientific theory of evolution in my book *Creative Evolution*. This theory explains the fossil gaps and the biological "arrow of time" required for evolution to move from simplicity to complexity—two important pieces of data that Darwinism and its offshoots cannot explain. In *Creative Evolution*, I also integrated ideas of Sri Aurobindo and Pierre Teilhard de Chardin about the future of humanity into a scientific approach. I drew on ideas developed by Rupert Sheldrake about morphogenetic fields (blueprints for biological form-making), bringing them under the umbrella of science within consciousness.

The biology establishment, however, has been very resistant to the influence of quantum physics, although—thanks to the empirical work on epigenetics and popular books by biologists Bruce Lipton, Mae Wan Ho, and others—quantum biology is gradually gaining ground.

In 2009, I set out to accelerate this paradigm shift by founding a movement called "quantum activism." My goal was to popularize the quantum worldview by bringing together a group of people dedicated to transforming themselves and their societies through practicing quantum principles. This has gained some attention, not only in America, but also in Brazil, Europe, India, and Japan, and even in the Middle East. In 2014, I went to Japan for an extensive dialog on the quantum worldview and quantum activism with erudite Japanese businessman and philosopher Masumi Hori. This book leans heavily on those dialogs. To this, I have added other interviews, notably one with author Eva Herr.

The result is a sort of Quantum Physics 101 for nonscientists. It contains elements from all of my previous work, and I hope it will inspire you to become a quantum activist. I hope to convince you that consciousness research and an understanding of the quantum worldview is the future of science. It is the foundation of a new paradigm that can lead us to the answer to everything.

CHAPTER 1

A Clash of Two Worldviews

People often ask me: If everything is not made of matter, then what is everything made of? And I say to them: Consciousness, *everything* is *made of consciousness*. But consciousness is such a wooly, nebulous concept! And this is where quantum physics breaks through with the answer we are looking for. For, in a quantum worldview, *everything* is wooly—even matter. Everything is a possibility before we experience it.

But if this is so straightforward, why do scientists debate about it? Scientists, in fact, still debate about all sorts of things: Is matter or consciousness the ground of everything? What does it mean to be human? Does God exist? While these are important questions, in our world of everyday affairs, what matters most is values. The biggest shortcoming of the materialist worldview is that it denigrates the archetypal values—love, truth, justice, beauty, goodness, abundance—and the meanings we derive from following these values. Yet, to a majority of the world's population, values like love remain important. Quantum physics, on the other hand, brings with it a new worldview that can put value and meaning back into our lives and provide answers to the questions of who we are and what it means to be human.

Someone once asked me if I found any similarity between quantum theory and the theory of the universe. And, in fact, the question is a good one. Quantum theory resulted from the observation of very tiny objects in the material world—the submicroscopic world. On the other hand, the theory of the universe is meant to explain a huge macro world. So how are these two related? In the quantum theory of consciousness, the large-scale aspects of the physical universe lose much of their interest. Modern cosmology has—thanks in large part to materialist science—avoided dealing with the internal world of consciousness. Thus it no longer seems to have any relationship to the real problems that occupy us all the time. But the concepts of modern cosmology are merely escapes—distractions not unlike the preoccupation of medieval Christian thinkers to determine how many angels can dance on the head of a pin.

I find it interesting that scientific materialists often posit their own exciting gods. All the exotic knowledge we now have of outer space has become a modern replacement for the gods of earlier religions—from the archetypes of Plato, to the angels of Christianity, to the more human Hindu gods like Siva. Instead, today, we call upon black holes and dark matter in an attempt to replace the archetypes and gods of earlier times. Modern science simply ignores consciousness and focuses instead on a concept of the universe that replaces archetypes and values with modern concepts like black holes and white holes, or dark matter and dark energy.

What we need to recognize is that science should always consist of three components. It must be founded on a theory. That theory must be verifiable by experimental data. And that theory must be useful. It must be applicable to human affairs. Whereas consciousness studies *are* now producing useful, experimentally verifiable, and technologically useful objects of inquiry, modern materialistic science increasingly involves itself with useless, non-verifiable objects of inquiry. Thus objects that were previously deemed more esoteric and less scientific are now becoming more useful and more scientific. At the same time, what used to be practical, down-to-earth science is becoming more abstract and less useful—more like old spiritual traditions. And spiritual traditions are becoming more like science.

What Is Consciousness?

Scientific materialists tend to treat consciousness as a linguistic assumption. There are subjects and predicates in our language, but science claims we can do without the subjects. As an example, they give the Hopi language, which has no subjects or predicates, only verbs, eliminating the need for consciousness except as a linguistic element. Without subjects—without consciousness—everything is matter and the manifestation of material interaction. This is the dominant worldview among scientists today.

If you ask a medical doctor to define consciousness, he will likely say, without batting an eye, that it is the opposite of being in a coma. A journalist told me her reaction to this kind of expert assertion: "Here we are, stuck with huge problems like global warming, economic breakdown, and political polarization—all because we can't come to a meeting of the minds over what a term like consciousness means. And we are not even aware that there is no meeting of the minds."

Of course, for many medical doctors, awareness and consciousness are synonymous, even after one hundred years of Freud. Doctors rarely read any psychoanalytic literature or, if they do, they do not accept much of it. Because how can the unconscious mind be validated if consciousness is not present in a patient who is in a coma? But consciousness never goes away. When we are unconscious—as in a coma—we may be unaware; we may have no experience of what is happening to us as subjects looking at objects. But we still have consciousness. What Freud really meant is that, although there is a distinction between awareness and unawareness, both are states of consciousness. In one state, we are aware of a subject-object split; we have an experience with two poles—the subject (the experiencer) and the object (the experienced). But in an unconscious state, we have no awareness of this split. Through psychoanalysis, we can explore how the mental processes taking place in the unconscious, of which we are not aware, are bothering us in our waking states of awareness. According to Freud, we should try to identify and understand these unconscious processes in order to function well mentally.

Consciousness is a fundamental aspect of our nature that is difficult to define—immediately, at least. We can become aware of some aspects and attributes of consciousness, but that's all we can do. Because consciousness ultimately, according to the quantum worldview, is the ground of all being, any definition that you give of it will fall short. Consciousness is everything there is. So any way you try to define it will fall short because the definition, in itself, is a phenomenon of consciousness, rather than the other way around.

Now let's return to that fundamental question with which we began: What is everything made of? Apart from psychoanalysis, is there any other compelling reason to choose between consciousness and matter to answer this question? Fortunately, today we can scientifically refute the materialist worldview. Theoretically, we can do so by demonstrating paradoxes—logical knots of thinking; experimentally, we can do so through anomalous data. Verbal quibbling has become unnecessary.

Material interaction has certain properties. One is that all interactions, all communications, occur through connections—signals that pass through space and time. Today, however, even undergraduates in physics can verify signal-less communication between submicroscopic quantum objects. And the work that some quantum physicists are doing proves conclusively that we cannot understand quantum physics without introducing causally potent consciousness into it—without introducing, not only consciousness, but also nonmaterial consciousness with causal power. We get paradoxes otherwise.

The causal power of consciousness—causation by conscious choice from potentiality into actuality—sounds much like the old Christian idea of *downward causation* by God. But that is not exactly true, although it is close enough to sound alarm bells in the cloistered minds of materialists. But here is the important thing. The new view of nonmaterial downward causation is that it involves nonlocal communication as opposed to communication with signals. Local communication goes through the locality to reach distant places, as for example when we communicate with sound; sound is a local signal. When we communicate without signals, as in mental telepathy, that is *nonlocal*.

With the concept of nonlocality, we have an experimentally verifiable consequence of a consciousness-based metaphysic. Material interactions behave locally and require signals. When consciousness interacts with the world, it requires no signals, only nonlocal communication. True, this type of communication seems subjective. But objective experiments since 1982 have shown that there are indeed nonlocal interactions in the world. Thus scientific materialism—based exclusively on material interactions—is ruled out experimentally. Instead, we can establish through experiment the idea that a new kind of nonmaterial interaction exists in the world. We have a new kind of causation—the causal ability of consciousness.

Signal-less Communication

In the last few centuries, materialist science has been very busy deciphering the mysteries of matter. And, in fact, it has developed technologies that were needed in order for our civilization to survive and move forward. These technologies have also produced some bad offshoots, however. We can't afford to put up with these negative consequences anymore—nor do we need to. Today's deepest scientific questions are about the large-scale structures of cosmology, and they're kind of useless. What is the practical use of studying black holes? We cannot verify them experimentally, and the research appears to have no purpose. So why spend so much time studying them?

On the other hand, we have problems galore in the world: global climate change, terrorism and violence, economic meltdowns and corporate greed, people without jobs or trapped in meaningless work, politicians monopolizing power and disempowering people, political polarization, the skyrocketing cost of conventional healthcare, education that reinforces dogmas and ideologies without setting living examples of the values they preach. Solving all of these problems will require a change in the global mindset, a change in our collective consciousness. So we need to develop a different approach—a move away from the current scientific paradigm to one that includes consciousness, one that has the ability to integrate the power of consciouseness into our everyday lives.

We must recognize that it is when called on to explain consciousness that the materialist model of the world completely fails as an explanatory principle. Objects, material objects, can only yield other conglomerates of material objects. All objects, when taken together, can never yield a subject—and that's what human consciousness is all about. We are all subjects looking at objects, looking at the world, formulating views about the world. Those who say those views all come from the dance of elementary particles at the basic level are just fooling themselves. They are ignoring the existence of meaning and values. They are denying that there is *causal efficacy at the level of human consciousness—at the very top level*. Without values, there can be no civilization. So our entire civilization is in danger when we take the word of scientific materialists that matter is the ground of all being. Quantum physics, by contrast, suggests a worldview in which consciousness, not matter, is the ground of all being. It suggests a world where meaning and value can be reintroduced into science as aspects of consciousness beyond matter. This is the new approach to science that our society needs.

Mainstream scientists have taken a very interesting approach to this critique—namely, benign neglect. They hope to discredit this new approach through their silence, depriving proponents like myself an opportunity to engage in debate. But while mainstream science has chosen to ignore the work of quantum activists, we have used the time to develop a new science uninterrupted by controversy. As a result, we now have a very good theory of consciousness based on quantum physics. Thanks to experimental researchers, we also have a lot of corroborating data.

Scientific materialism relies on a concept called "dualism"—the idea that anything nonmaterial must exist as a separate object—as its main justification for denying the role of consciousness and all other "internal" experiences. Dualism begs the question of how material and nonmaterial objects interact. Think about it. If matter and nonmatter have nothing in common, they need a mediator, a signal, to interact—something to "connect" them. This has proven a tough nut to crack for supporters of nonmaterial beings. Quantum physics' answer is signal-less communication—nonlocality, in technical jargon. Signal-less communication is impossible in space and time; so it must use another domain of reality outside of space and time. According to quantum physics, this is the domain of potentiality. If this is true—and experiments say that it is—all materialist arguments against dualism go away, returning value and meaning to spirituality, religion, the arts and humanities, and indeed to consciousness itself. And if dualism goes away, nonmaterial objects can communicate with material objects and with other varieties of nonmaterial objects, because no signal is needed for them to communicate through the domain of potentiality—also known as consciousness.

Quantum physics forces us to conclude that the domain of potentiality is really consciousness itself. Moreover, it shows us that the communication between what seem to be two separate objects—mind and matter—is mediated by consciousness. This is the essence of the quantum paradigm.

Sometimes materialists try to discredit the idea that quantum physics, quantum nonlocality, can affect phenomena at the macro level of our experience. But we now have support from many experiments in a variety of fields—physics, biology, psychology, and medicine—that suggest that there is a nonlocal domain, even at the macro level. These experiments support the claim that signal-less communication really does happen, not only in the microscopic world, but also in the macro world of matter and human experience. As the underpinnings of their arguments disappear, more and more mainstream scientists are coming around to the quantum point of view. Although most still do not engage with the "weird" aspects of quantum physics (like nonlocality), those who do are becoming more amenable to scholarly discussion about the theory.

Parapsychologist Dean Radin supports the new quantum worldview and has undertaken some interesting experiments using a random-number generator to support it. A random-number generator converts random events of radioactive decay into random arrays of zeros and ones with the aid of a computer. Radin took these random-number generators to locations where people were meditating. He found that, in the presence of the meditators, the behavior of random number generators became significantly more nonrandom than was statistically expected. Radin suggested that the random-number generator should deviate maximally from randomness in the presence of coherent intentions. And he verified this idea, not only with people in a meditation setting, but also with people watching the Super Bowl. In those situations, Radin found that intention indeed caused deviation from randomness.

In situations where people were scatterbrained and not particularly intending anything, the random-number generators behaved normally. For example, in the corporate boardroom or at a university faculty meeting, random-number generators really generated random arrays of zeros and ones. In the meditation halls, they didn't. This supports the new view of quantum physics that conscious intention can affect outcomes. It shows the presence of conscious choice, which of course—as Gregory Bateson said a long time ago—is the opposite of randomness. The antagonists of the quantum worldview have to come to terms with experimental data like this.

Polarization and Integration

In our world today, we don't need polarization; we need integration. Although perhaps not as pronounced elsewhere, in America, polarization between science and religion has absolutely paralyzed the political process. How does the polarization between science and religion infect politics? Very simple.

On the one hand, there are people who want values, who are very afraid that scientific materialism will overwhelm their entire society and leave them with no moral compass. They'd rather live without science than without their values. Then there are the gleeful materialists who justify a hedonist lifestyle with scientific materialism and existential philosophy. Conservatives who once stood for solid moral integrity and character have sided with the archaic worldview of fundamentalist Christianity and become anti-science rather than pro-values. They risk taking us back to a place in time where religious and political elites dictated morality. At the same time, originally open-minded and creative liberals, who supported science because of its promise to free us from all dogma, have come to rely on scientific materialism, itself a dogma, to support a different kind of elitism in which knowledge and information are power. People who have that power and monopolize it are the new elite.

But science should be dogma-free. Science is a methodology. First you have a theory; then you have experimental data; then you apply that theory and data. But how can we implement this methodology if dogma gets in the way? On the one hand, you have the divisive incomplete theory of evolution—Darwinism. On the other hand, you have the creationists, Christian fundamentalists who use archaic Biblical ideas to thwart science.

Both sides are caught up in a dogmatic battle that prevents science from moving forward. And people are suffering because of it.

Mainstream science has tried to debunk and suppress the data that supports nonlocality in our macro-experience. They label these phenomena "paranormal" and refute the quantum-based theory of consciousness through sophistry. Quantum activists claim that understanding quantum physics is impossible without introducing consciousness into the mix. But materialists just cite another half a dozen plausible ways to make the paradoxes of quantum physics go away. Then they treat our consciousness-based theory as just one more in a long list of proposed solutions. Never mind that, on closer examination, all these other seemingly plausible solutions are unverifiable, whereas the consciousness-based solution has already satisfied the verifiability criterion. The whole character of science is changing under the materialist aegis. It is becoming what I jokingly call fact-free. Many famous scientists have put forth theories that have never been, and probably will never be, verified.

So how do we resolve this battle of dogmas? The solution is simple: quantum physics and the quantum worldview. Quantum physics has been with us for almost one hundred years. We have explored it and spent an enormous amount of time trying to understand its message. From the very start, it was clear that the Newtonian worldview, scientific materialism, would not stand up to the findings of quantum physics. Yet we still have not succeeded in resolving the dilemma. After World War II, when the power of science shifted from Europe, which is more philosophy-centered, to pragmatic and practical-minded America, the message of quantum physics got lost in favor of the more seemingly practical philosophy of scientific materialism.

Experimental Metaphysics

I was quite young and still pursuing traditional physics when quantum physics began to make itself felt again in the wider culture. I remember quite a bit of excitement in the 1970s when the book *The Tao of Physics*—and the slogan, We Create Our Own Reality—burst onto the scene. In truth, we even used to have at least one annual conference on the philosophical issues of quantum physics. But the philosophical issues were never resolved because of a lack of experimental data. Then experimental verification of quantum weirdness began coming in the 1980s, and we came back to the philosophical questions with gusto. And that's when we realized that some of the deeper paradoxes of the quantum worldview—some of its logical "weirdness"—would never be resolved when looked at through the old lens of scientific materialism.

Rather, the solution would require a new metaphysics, one that was also experimentally verifiable. Philosopher Albert Shimony calls this new development "experimental metaphysics." In the new metaphysics, consciousness is the ground of being. This is a metaphysical idea, but one that can be put to experimental test. And the test is very simple. If matter is the ground of being, there cannot be any such thing as signal-less communication—nonlocality. Whereas, if consciousness is the ground of all being, signal-less communication must occur, even in the macro world of our experience. And proof of this is now abundant.

Let's be clear, however. I say that scientific materialism is a dogma because of its belief that matter is everything. But by that logic, isn't the belief that consciousness is everything also a dogma? Well, it would be—except for one important difference. The quantum worldview is inclusive. It does not exclude the possibility or efficacy of the material world. It puts both consciousness and matter—God and the world, if you will—on an equal footing.

So we have to change the way we look at things. Modern science has given scientific explanations for some horrible "evil" truths about us human beings—we have instinctual negative emotional brain circuits; we hate; we are violent, competitive, jealous, and angry because we have evolved that way. This is the negativity we have to offset; we have to build positive emotional brain circuits. But, according to scientific materialism, this is not possible. Scientific materialism denies the existence of values; it denies the validity of our intuitive experiences that point us to values. It denies any creativity that can enable us to build positive emotional brain circuits.

And yet we have known for millennia that the change in us, in our evolutionary future, has to proceed in such a way that we become more loving, more good to our neighbors, more appreciative of beauty, more capable of giving justice. The movement of consciousness demands it. These are the ways we want to change to offset our evolutionary shortcomings. We want to make the Platonic archetypes—values—manifest in ourselves, to incorporate them into our brain circuits. This goal may seem "unscientific" and may smack of purposiveness to the scientific materialist—but so what? The new science, as we will see, makes room for purpose as a way to drive change.

And of course, where there is a goal, there is a way to reach it! All we need do is to follow up intuition with creativity, properly understood. Thanks to the quantum worldview, we know that creativity is possible and that it will help. For the first time in human history, we have a clear purpose that is not a world-negating purpose—the evolution of the world itself to positivity. Most spiritual traditions have a tendency to think of the material world as illusion. This is not true of the quantum worldview, which allows us to maintain the positive elements of the spiritual traditions, but leave behind the world-negating aspects completely. The world is lawful; the world has order; it is important.

So the quantum worldview lets us integrate the best of scientific materialism—the importance of the world—with the best of spiritual traditions—the importance of wholeness. Into this paradigm, we can integrate our reliance on science for technology and our reliance on spiritual traditions for meaning, values, and the energies of love. This is the soul-satisfying goal of quantum activism—to change ourselves and our society according to quantum principles. By changing ourselves, we arrive at personal growth and satisfaction and meaning; by revolutionizing our social systems—politics, economics, health and healing, education, religion, and ecology—all of which are currently in crisis, we save civilization.. Thus the quantum worldview and quantum activism can literally help save us from ourselves.

CHAPTER 2

Consciousness and the Science of Experience

I introduced the movement of quantum activism because the scientific establishment and the media are keeping the more recent paradigm-shifting aspects of the quantum a secret. Activism is needed to spread the word that we don't have to remain polarized between science and religion any more, that the integration of these worldviews has already been achieved. I also wanted to demonstrate the causal efficacy of the new integrative worldview in transforming society. You can see the full story on quantum activism in *The Quantum Activist,* a 2009 documentary about my work.

Quantum activists try to transform both themselves and their society using the quantum worldview, a process I describe in my book *How Quantum Activism Can Save Civilization* (2011). In the Newtonian worldview, you are a determined machine—although sophisticated, to be sure. Through Darwinian evolution, you have equipped yourself with a lot of sophisticated programs that make you appear conscious and free to choose your destiny. In the ultimate reckoning, however, it is chance and the built-in need to survive that determines your behavior.

To envision a spiritually transformed destiny for humanity in this worldview would be outrageous.

On the other hand, quantum physics equips you with real causal power—*downward causation*—the power to choose. In Newtonian physics, objects are determined "things" made of matter, whose movements are determined by material interactions among the base-level objects called elementary particles. This is thus a world of *upward causation,* in which material causes rise upward from elementary particles to more and more complex matter. In quantum physics, however, objects are not determined things. They are quantum possibilities from which consciousness can choose. This world of conscious choice is a world of downward causation. In a quantum world, you *can* choose your reality, your spiritual destiny.

Say you are hungry and want a grilled cheese sandwich. You can say or think that intention until you are blue in the face, but nothing happens. Of course not. You have to start with a possibility. Is there a grilled cheese sandwich in front of you? Let's say there is. Well, quantum physics says that, when you are not looking at that sandwich, it becomes a possibility of a grilled cheese sandwich. Then when you look back, the sandwich manifests into actuality and you can eat it.

But aren't there plenty of grilled cheese sandwiches in your city? Why can't you manifest one of them just by choosing one of them, thinking one of them? And here enters the question of probability. Because of low probability, you would have to wait a very long time before you got your sandwich by just wishfully thinking about one of them.

In fact, there are many subtle aspects to the secret of creating reality. When I first started on consciousness research back in the 1970s, I was naïve about these subtleties. What made the difference for me, however, was that I am a quantum physicist.

Once I understood the subtleties of quantum physics, I used them to teach me about myself and to change myself if I could. And that made all the difference. I hope that, by exploring my work, you will begin to understand the subtleties of both quantum physics and consciousness, and use them to change the way you live and the world you live in.

We face a crisis in our personal lives today. We are confused about the meaning and purpose of life; we feel lost in our quest for love; we have become information junkies and given up on real satisfaction. We are also facing a crisis in our society—polarization between religion and science that is stopping political progress in many countries. We need to resolve this crisis. Our prevailing scientific paradigm is not working. But we already have an alternative paradigm available—one that recognizes the role of consciousness in science. If this paradigm is allowed to replace the failed paradigm, it can integrate the disparate worldviews of religion and science and restore meaning and value to our lives. We have to move to that new paradigm. We need quantum activism.

Quantum activism awakens us to the reality of the primacy of consciousness. The new movement is a way out of the crises that we have created—the crisis of confidence in our science and our worldviews, and the crisis in the way that we look at ourselves. The quantum activist movement is now present on five continents and it is gaining traction every day.

In the 1970s and 1980s, I found that quantum paradoxes could not be resolved without introducing consciousness into the equation. That's when I became involved in consciousness research. Before that, I must confess, I was fully a materialist. I believed that everything was made of atoms and elementary particles. I believed that there was nothing but matter, that consciousness was a physical phenomenon of the brain, and that spirituality was just humbug. If you try to analyze quantum physics from this standpoint, you fall flat on your face. You simply generate paradoxes that cannot be resolved.

After worrying over these paradoxes for a few years, a sudden insight came to me while I was talking to a mystic—an insight that turned my thinking around completely. It came to me that, if consciousness is the ground of being instead of matter, the paradoxes of quantum physics can all be resolved. Moreover, I realized that there is nothing contradictory about building a science on this new metaphysics. It was this revelation that I found most illuminating.

Until that moment, I had assumed, along with many others, that one simply cannot "do" science without the materialist assumption of total objectivity. But what if that were not true? What if, until now, we have been providing scientific explanations for only half of our reality, while ignoring the other, subjective half because we subscribe to materialist dogma? And what if we were to give up that materialist dogma in favor of a metaphysics of the primacy of consciousness? Then this other, neglected half of reality—our subjective experiences, our consciousness, love, spirituality, God, meaning, feeling, life, death, even mythology—can be included within the purview of science. Moreover, many of the controversies in many scientific fields can be resolved. This is what brought me around to the quantum worldview.

What Is a Quantum?

Now, let's turn to the more basic question: *What is a quantum?* A quantum is a discrete quantity first used with this connotation by physicist Max Planck to denote the idea that energy exchange between bodies can take place only in terms of discrete quanta—one quantum, two quanta, etc.—but never half a quantum. You can think of an elementary particle as an irreducible quantum of matter. A photon is a quantum of light. But you can never have half a photon. Our monetary system is a quantum system. You can make change for a dollar; you can make change for a quarter; but (unless you are a bank) you can never make change for a penny. This quantum monetary system is arbitrary, however; it is not a physical law. That's why banks can violate it.

The word "quantum" packs a lot of power. Physicist Niels Bohr theorized that, when an electron jumps from one atomic orbit to another, it does not go through the intervening space. Its movement is discontinuous. Bohr called this a "quantum leap," a phrase that has since come to be associated with discontinuous, as opposed to continuous, movement.

But there is much more to a quantum than this. That light consists of quanta—photons—is backed up by experimental data, of course. But that is only half of the nature of light. Light is also a wave, and this, too, is backed up by experimental data. The difficulty arises in the differences between particles and waves. Particles are localized objects. They move in trajectories and can only be in one place at a time. Waves, on the other hand, are always dispersing, spreading out, displaying their capacity to be in more than one place at the same time. Can the same object, then, be both particle and wave? Logically, no. And therein lies the paradox—a paradox that applies to elementary particles like electrons as well. They all are both waves and particles.

The new version of the complementarity principle we discussed earlier allows us to think about this in a paradox-free way: the waves are waves first in a domain outside of space and time; when we measure them, they appear as particles in space and time. Behold! Quantum objects are fundamentally waves of possibility and the domain they inhabit is called the domain of potentiality.

Today, when most scientists think of nature—of the material world—they are thinking of space and time. Anything that refers to something outside of space and time they call supernatural. But quantum physics—and every physicist today swears by it—says that nature (reality) has two domains, one within and one outside of space and time. And the domain outside of space and time is experimentally discernible. Communication via this domain is instantaneous and signal-less. This sounds pretty radical, doesn't it?

But who else talks like this? Mystics. They talk about heaven and earth—two domains of reality. Heaven is the domain of God and godly qualities and perfection; earth—the domain of ordinary matter and humans—is looked upon as a realm of imperfection. So quantum physics tantalizes us with the possibility of an integrative way of thinking about both science and mysticism, whose offshoot is religion. And that is where I began—with a basic motivation to integrate science and spirituality, a desire to end the age-old feud between science and religion by bringing them together in a quantum worldview. I made this point in my very first book on the subject, *The Self-Aware Universe* (1993), then elaborated on the idea in *The Visionary Window* (2000). Finally, in *God Is Not Dead* (2008), I demonstrated the integration of science and religion using empirical data. And twenty years later, I am still working to bring about this synthesis through quantum activism.

The quantum worldview can also integrate science into the arts and humanities. We have lost faith in the arts and humanities and, with that, we have just about given up the pursuit of meaning in favor of trivial pursuits. The romantic poet Samuel Coleridge, writing about how he got the idea for his masterpiece *Kubla Khan*, said:

> What if you slept and what if in your sleep you dreamed? And what if in your dream you went to heaven and there plucked a strange and beautiful flower? And what if, when you awoke, you had the flower in your hand?

This brings to mind psychoanalysts (followers of Freud) and depth psychologists (followers of Jung), who also talk about two domains—the unconscious and the conscious. Can quantum physics give us a paradigmatic umbrella for both physics and psychology? Freud's protégé, the visionary Carl Jung, said that, sooner or later, psychology and quantum physics would come together. He was right.

So the word "quantum" indeed packs a lot of power. Moreover, the quantum worldview is always integrative and inclusive. Here, we are concentrating on its integrative aspects: how quantum physics can unify science and spirituality; how it can unite cause and purpose; how it can bring back meaning and revitalize the arts and humanities; how it can integrate science and psychology; and how it can give us a science of love. Quantum activists take the commitment a little further; they use quantum principles to change themselves and move their society toward integration and inclusivity.

The Physics of Possibility

Quantum physics is the physics of possibility. The waves of wave-particle duality are waves of possibility. In the quantum

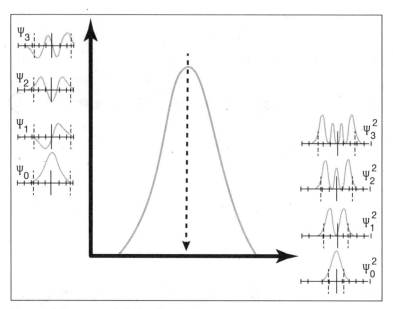

Figure 1. Electron probability distribution curve

Conc. worldview, *consciousness* chooses actuality out of quantum possibilities. That's how we create reality, including ourselves. This power of choice is called downward causation.

To understand the idea of quantum possibility, consider how an electron behaves when it is released so slowly that it is practically at rest in the middle of an imaginary room. In Newtonian physics, the electron would stay right where it was released, forever, if you ignore gravity. But this is not so in quantum physics. In quantum physics, this same electron behaves like a wave and spreads.

When you throw a pebble into a pool of water, ripples in the water—waves—spread out from where the pebble lands. In quantum physics, even an electron at rest spreads out in a similar fashion, but in three dimensions. And this happens so rapidly that, in no time (literally), it is all over our imaginary room. Now imagine a three-dimensional grid of Geiger counters—those devices that go tick-tick-tick when electrons fall on them—distributed throughout the room. Will all the Geiger counters start ticking as the electron wave spreads throughout the room? No. In any one given experiment, only one of the Geiger counters will tick. In another identical experiment, another Geiger counter elsewhere in the room will tick. And if we do many identical experiments, we will generate a bell curve that shows the probability of where that electron is likely to be found at any given moment.

So does the electron actually exist simultaeously in locations all over the room? Yes. That is what quantum mathematics tells us. But, in order to make sense of what we observe, we must also conclude that the electron is at many places at the same time *only in possibility*. And this is the essence of the physics of possibility.

The predictive power of quantum physics comes from these bell curves of probability that we can calculate using quantum mathematics. Along with the possibilities of quantum physics come associated probabilities that allow us to answer to the question: What is an electron's position on average for a large number of measurements? From the curve shown in Figure 1, we can predict the probability of where the electron will be in our imaginary room. In physics and chemistry, we are always dealing with almost inconceivable numbers of quantum objects. So the prediction that a spoonful of sugar helps the medicine go down only holds true because the probability prediction that guarantees it is true for a very large number of experiments.

For a single object or a single event, however, probabilities don't help us. And quantum mathematics has no answer for where the electron will be found in a single experiment. So we posit that consciousness chooses the electron's position whenever an observer is present. This is called the "observer effect."

The Observer Effect

The observer effect, which I discussed in the movie *What the Bleep Do We Know?*, states simply that a possibility wave of any given object or event changes into actuality only when an observer looks at it. One way to express this is to say that the wave changes into a particle when observed. Physicists call this change into actuality "collapse," because early quantum physicists thought that waves of possibility collapsed like an umbrella going through space, taking an amount of time to do so. But now we know that this is not the case. We know that this collapse occurs instataneously, nonlocally.

But why doesn't the Geiger counter itself transform the waves into particles merely by registering them? That seems to be what common sense dictates. The answer is two-fold. First, without observation, we could never verify that. Verification requires us to look at the Geiger counter or hear its ticking—to observe it. Second, we know that the Geiger counter is made up of molecules further reducible all the way down to elementary particles, and all of these objects obey quantum physics. Thus, even the Geiger counter must obey the laws of quantum physics and respond as a possibility wave as it interacts with the electron's possibility wave.

But isn't the observer's brain also made of molecules that are reducible to elementary particles, and thus also subject to the laws of quantum physics? What's the difference between the brain and the Geiger counter?

True, the brain can be expected to follow the laws of quantum physics. But don't forget that, somehow, in the presence of the observer's brain, the collapse of the wave into the particle *does* happen; we *do* hear the tick of the Geiger counter. So not only must the brain be special, but the observer must also be something *more* than the brain. And that something is consciousness.

The word "consciousness" is derived from two Latin words—*cum,* meaning "with," and *scire,* meaning "to know." Consciousness is thus the vehicle by which we know things. In our experiment with the electron in the room, we knew only possibilities and probabilities about the object before we heard the Geiger counter's tick; our knowledge about the object was vague. Once we heard the tick, however, we knew exactly where the electron was. The measurement increased our knowledge of the electron. And the vehicle with which we knew that is our consciousness.

Consider this. Suppose that I am in a store looking at beautiful ceramic cups. I inadvertently push one off the table and it breaks, but I am not looking at it. Did I only break the possibility of a cup? If so, why should I have to pay for it? But I heard the cup break. And even if I were deaf, the store clerk will have heard the cup break and he will come running and charge me! So when we talk about "observing," we mean observing through all sensory ways of knowing, not just looking.

The observer effect requires that the observer interact with the object in some way that involves nonmatter, because material interactions, according to mathematician John von Neumann's famous theorem, can only convert possibility waves into other possibility waves, never actualities. This nonmatter is the observer's consciousness. But if consciousness is that which knows—what we conventionally call the subject—we arrive at another paradox. Obviously, the subject does not exist without the brain. But without collapse—without moving from possibility to actuality—we only have a possible brain. The existence of the brain requires collapse; collapse requires the presence of the brain. There is a causal circularity here, a paradox of logic that is part of the "quantum measurement paradox."

But quantum physics itself tells us what consciousness must be in order to avoid all paradoxes of thinking about it. Consciousness must be the ground of all being; matter consists of possibilities of consciousness itself. Since consciousness is choosing from itself, this assertion avoids the key paradox of dualism—how consciousness can interact with a material object without a signal. Quantum physics gives a simple, but radical, answer: There is no signal. Thus there is no need to posit interaction between separate objects. The object is one with consciousness. When you communicate with yourself, you don't need a signal. This signal-less communication is called communication through *quantum nonlocality*.

Thus consciousness is not a phenomenon of the brain. In the quantum view, consciousness is the ground of all being and the brain is a phenomenon of consciousness. A very common tendency is to think of consciousness as an object—a phenomenon of the brain—eventually reducible down to elementary particles of matter (see Figure 2). But a conscious experience always consists of two poles: subject and object, experiencer and experienced. So how can the subject come from the brain, if the brain is merely an object made of smaller objects all the way down to elementary particles? Consciousness is more than an object; it also contains the subject. Therefore, the resolution of the quantum measurement paradox is that, in a quantum measurement, the brain makes a representation of the subject-potentiality of consciousness as consciousness identifies with it.

But why is the brain so special? Why does consciousness identify with the brain, but not with a Geiger counter—or a rock for that matter? The answer is crucial. There is a circular

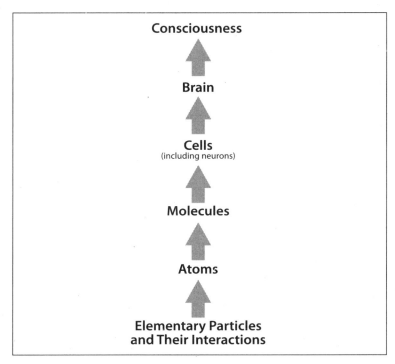

Figure 2. If consciousness were an object, there would be no subject to look at objects.

relationship among the components of the brain that gives us self-identity with it.

Consider the sentence "I am a liar." The sentence is paradoxical; it has circular logic—what physicists technically call a "tangled hierarchy." If I am lying, I am telling the truth; if I am telling the truth, I am lying. No matter how many times you repeat it, you'll never escape the circularity. The brain has that same kind of paradoxical relationship between its perception apparatus and its memory apparatus. Perception requires memory; memory requires perception. In the process of quantum measurement, which involves both the perception apparatus and the memory apparatus of the brain, consciousness gets caught in its own circularity; it identifies with the brain. The brain thus becomes the subject pole of an experience. In this way, consciousness splits itself into a subject (that experiences) and an object (that is experienced). This is an act of quantum creativity.

Quantum Creativity

Quantum creativity is not a mechanical process. It requires access to higher consciousness. Diehard Darwinians will object here that the brain is itself a product of evolution—an essentially mechanical, linear process. That's true. But evolution does not have to involve material interactions alone, as Darwin theorized. In fact, if consciousness is the ground of being, it makes perfect sense that consciousness has a role to play in evolution.

Now that you understand a little more about the role of creative consciousness, let's go back to the New Age dictum: We choose our own reality. Surely, if we can choose reality at will, we can choose a car or a beautiful house. . . .

Well, no, you can't.

The error we make is that we tend to think that we choose with our own individual consciousness. This results in another paradox. Consider this: Who gets to choose whether a dichotomous quantum traffic light is red or green when there are two observers arriving at the light from perpendicular directions with different motives? Both observers will want the light to be green on their side (see Figure 3). But who gets to choose? The solution is two-fold. Neither gets to choose in his or her individual ego; and both get to choose as nonlocal consciousness. Consciousness, the ground of being and the source of our choice, is nonlocal and objective. Some may call this God, but we can just as effectively call it quantum consciousness.

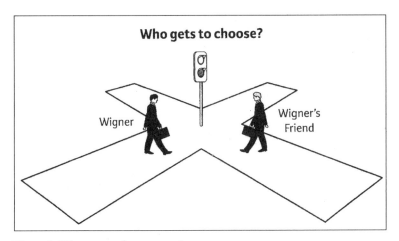

Figure 3. Who gets to choose green?

Quantum physics, the physics of possibilities, thus helps us wake up to our higher consciousness and choose the actual events of our experience from a realm of possibility. We don't choose in our individual conditioned egos, but rather from a higher consciousness where we are one with others. It thus empowers us to create our own reality. And the more we use this power of quantum creativity, the more we realize that creativity is a cooperative process, not a competitive one. That is *transformation*.

Nonlocal Communication

In our discussion so far, we have identified the quintessential quantum principles, the most important concepts in a quantum activist's tool kit: downward causation, nonlocality, discontinuity, and tangled hierarchy. These are the principles of quantum physics that can transform us when we learn to apply them in our own lives. These are the tools with which the quantum activist works. These are the vehicles that can bring about quantum change.

Nonlocality is signal-less communication that occurs through consciousness, the domain of potentiality. It has been objectively verified even at the macro level of our experience. At the level of neurophysiology, there is evidence of transferred potential; electrical activity can be transferred from one brain to another without an electrical connection. There is also somewhat subjective evidence of this in fields like telepathy. And there are new experiments—distant-viewing experiments, for instance—that are being done with a good amount of what we call weak objectivity. These depend on subjective experiences, but can be verified across a large number of subjects. Without this concept of weak objectivity for validation, even cognitive psychology would be very difficult to justify as science.

Nonlocality is a difficult concept for the layman to grasp. We tend to assume that information has to be carried in some way into a field or through a frequency. But this assumption requires that information be encoded within these frequencies or fields in the form of time-space pulse and voltage—for example, pulses of energy tapping out a message, rat-a-tat-tat-rat-a-tat-tat. But this, in turn, requires that something in the mind be able to decode this message. Again, the danger is that we tend to think in terms of something actually moving by means of a signal, which in turn requires that the brain be able to interpret the messages that consciousness is delivering as modulated waves.

The answer is that the brain does no such thing. In the quantum model of how this transmission takes place, one brain does not emit electromagnetic waves that are modulated and received by another brain. While materialists will immediately say that this is indeed the case, there is no evidence for it. The problem for quantum science is that we cannot rule out the phenomenon of telepathy itself by this kind of argument, because telepathy can be proven to take place without transmission via electromagnetic waves simply by putting the subjects in electromagnetically impervious chambers. In these experiments, information moves from one brain to another without the help of electromagnetic waves. So how this does happen?

In the consciousness model, it's a subtle process grounded in the potentiality that is present in the cosmic consciousness that is one for both observers, for both telepaths. So communication occurs when one observer thinks something—or more accurately, chooses something—from the meaning domain of potentiality, and the other observer, because he or she is "correlated" with the first observer, chooses the same thing. Thus both choose objects identical or almost identical in meaning from the spectrum of possibilities that they are processing in their own unconscious domain of potentiality. The brain of the second observer makes a representation of that mental meaning, which then seems to have been transferred from the first observer's brain. This process is now being verified by new experiments called "transferred-potential experiments," in which brain potential from one subject's brain is transferred to another subject's brain without electromagnetic signals.

In 1993, neurophysiologist Jacobo Grinberg, at the University of Mexico, was able to demonstrate quantum nonlocal communication between two brains. To this end, he first correlated the two subjects of his experiment by having them meditate together with the intention of direct (signal-less, nonlocal) communication. After twenty minutes, the subjects were separated (while still continuing their unifying intention) and placed in individual Faraday cages (electromagnetically impervious chambers). Each subject's brain was wired to an electroencephalogram (EEG) machine. One subject was exposed to a series of light flashes producing electrical activity in the brain that was recorded by the EEG machine. From this, an "evoked potential" was extracted with the help of a computer after removing the brain noise. The evoked potential was somehow found to be transferred to the second subject's brain, as indicated by the EEG of this subject after removal of the brain noise. The second subject gave a transferred potential similar to the evoked potential of the first subject in phase and strength (see Figure 4). Control subjects (those who did not meditate together or were unable to hold the intention for signal-less communication during the duration of the experiment) did not show any transferred potential.

Grinberg's experiment demonstrates the nonlocality of brain-to-brain communication, and something even more important—the nonlocality of quantum consciousness. How else can we explain how the forced choice of the evoked response in one subject's brain can lead to the free choice of an (almost) identical response in the correlated subject's brain? This experiment has been replicated perhaps as many as two dozen times. (See, for example, research by Leana Standish and her collaborators, or that of IONS researcher Dean Radin.) Perhaps the most important aspect of Grinberg's experiment is the power of our intention. Grinberg's subjects intended that their potential nonlocal connection would manifest in demonstrable actuality. Control subjects who could not hold the intention were unable to manifest a transferred potential.

Experiments like this are revolutionizing our attitude toward nonlocal consciousness. If the electric potential (electric activity) in one brain can be transferred to another brain without an electrical connection, without the presence of any electromagnetic wave, then how can we deny there is a subtle nonlocal connection between the two brains? And it is this interconnectedness that we call consciousness.

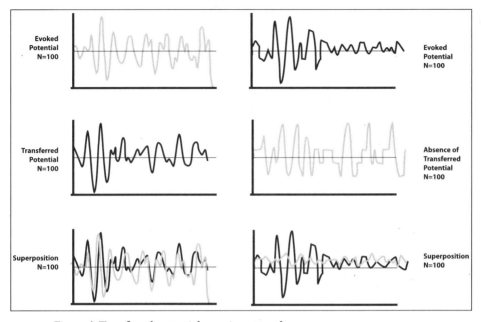

Figure 4. Transferred-potential experiment results.

Tangled Hierarchy and Discontinuity

Now let's look at another of the quantum activist's tools—tangled hierarchy. Tangled hierarchy is the most difficult of these tools to understand. Everyone understands simple hierarchy in a social sense—monarchy, patriarchy, oligarchy, etc.—which consists of one level of a group controlling the rest. The simple hierarchy that materialist science gives us is that elementary particles make atoms; atoms make molecules; molecules make bigger objects. This simple hierarchy is based on upward causation, and it's quite an accurate description of inanimate material objects. But when it comes to the relationship of subjects, relationships among people, we tend to rebel against simple hierarchies. We have moved generally away from monarchy—a type of simple hierarchy—toward democracy, which encourages more causal equality among people. Those monarchs who remain are, for the most part, benign figureheads, not ruling authorities. Of course, we are still subject to simple hierarchy in our societies and in our political structures. But if we want social change, we will likely have to turn to a more "tangled" relationship between voters and their representatives.

Discontinuity, the third of the quantum activist's tools, is familiar to many of us in the form of creative experiences that come as a surprise—those "aha" moments we have all experienced. Surprise is, in fact, the signature of discontinuity. Of course, creative experiences are, at best, weakly objective. But objective evidence of discontinuity shows up in significant ways even in biological evolution—a fact often overlooked by scientific materialism. If you say that evolution is all just a movement of matter, you capture only the slow and continuous part of evolution. But there is also another, more abrupt kind of movement in evolution that biologists Niles Eldredge and Steven Gould discovered in the 1970s. They aptly called these evolutionary leaps the "punctuated marks" in the continuous prose of Darwinian evolution. Biologists of scientific materialism cannot explain these punctuated marks. But biologists of quantum worldview can.

In the quantum model, these punctuated marks are simply quantum leaps of creativity, biological creativity (see my 2008 book, *Creative Evolution*). If we take this view, then evolution is an evolution of consciousness. Darwinism limits the pace and scope of evolution by insisting on continuity. But to explain evolution at the macro level—from one macro species to another, where the evolution involves a new organ, for example—we must invoke these quantum leaps. Moreover, the data supports this view. The fossil gaps in evolution are well known. Scientists unwilling to move outside of Darwin's continuous model have found a few intermediaries to fill those gaps, but they will need to find literally thousands upon thousands more to solve this central anomaly of Darwinism.

The quantum case for discontinuity, on the other hand, resolves this anomaly and removes the limits from the Darwinian model. Through its reliance on nonlocality, tangled hierarchy, and discontinuity, it opens the door to creativity on all levels—biological, material, cultural, and psychological. It presents us with an evolutionary process that can really move us forward.

Exclusion or Inclusion?

Scientific materialism has become a dogma—and a very exclusive dogma at that. It excludes spirituality, the arts, and major aspects of psychology—for example, the unconscious in psychoanalysis and depth psychology. It excludes many of the alternative methods of healing that people have found useful for centuries—for example, acupuncture, the Eastern tradition of Ayurveda, and the relatively recently discovered homeopathy. All of this is, according to mainstream science, just humbug that should be ignored. Ironically, recent studies show that 70 percent of healing by pharmaceutical drugs is due to the placebo effect, which is a mind-body effect, and therefore excluded by scientific materialism. Similarly, scientific materialists ignore the whole new paradigm of consciousness that the quantum worldview posits.

So how do we get beyond this? How do we arrive at a philosophy of inclusion in science, rather than exclusion? The only way I can think of is through activism—through a conscious application and advocacy of quantum principles that can give us solid guidance in how to effect change through choice. People often complain that activists try to change the world, but never change themselves. In quantum physics, changing ourselves is easy, because we now know how—through creativity. Creativity satisfies. Everybody wants to be creative somehow. That's how we each become unique, as well as how we can cooperate. So in this way, quantum activism inspires us to change ourselves as we try to change the world.

The goal of quantum activism is a life in which there is more wholeness because there is more integration. As a quantum activist, I can truly say that I can live, to a certain extent, with awareness. I can bring goodness to my relationships with other people. I can have love in my intimate relationships. I can be just in my dealings with people. I can discern right from wrong. I can appreciate beauty and harmony.

As a quantum activist, I can integrate the mystic, the poet, and the scientist within myself. I can combine within myself science and spirituality, and art and humanism, and all the other positive potential that makes us human. For me personally, my ideal is to integrate the potentialities of the great mystic poet Rabindranath Tagore and those of the great scientist Albert Einstein.

I know this sounds extremely ambitious, and I'm not sure how long it will take humanity to realize this goal. But I do know that, whether or not you call yourself a quantum activist, if you are working toward integration and inclusion, and embracing a creative approach to change, then you are, indeed, a quantum activist who understands, implicitly or explicitly, the essence of the quantum worldview.

CHAPTER 3

The Physics of the Subtle

We all potentially have the power of downward causation—the power to choose from an array of possibilities. But what can we do with it? To begin with, we have to recognize that this power to choose from what is possible is very limited in the physical realm, but quite unlimited in what is called the subtle realm in many spiritual traditions.

What is the subtle realm? It is comprised of what we experience internally, as opposed to matter, which we experience externally. You can think of matter as gross, fixed, and semi-permanent. But the subtle realm is forever changing. How can we make these spiritual concepts more scientific? By realizing that if matter exists as possibilities within consciousness, why not the subtle? By adopting the model of quantum physics as a way to resolve the so-called mind-body dualism of materialist science.

Bodies of Consciousness

Many spiritual traditions talk about subtle bodies of consciousness other than the physical. They incorporate the vital, the mental, and the supramental—archetypes like love, beauty, truth, justice, and goodness—into their belief systems. They often portray the physical and the subtle as embedded in a fifth body, the wholeness of consciousness that is looked upon as the ground of being.

Psychologist Carl Jung described four categories of personalities: sensing, feeling, thinking, and intuitive. Seen in the context of the ground of consciousness, these personality types can delineate four different worlds of possibilities: material possibilities that we sense when we make them actual; vital possibilities that we feel; mental possibilities that we think; and supramental possibilities that we intuit. When we actualize possibilities by choosing among them, we create an experience (see Figure 5).

We have an existence, a body, in each of these worlds. These bodies do not directly interact; consciousness mediates their interaction nonlocally. In this way, mind-body dualism becomes meaningless and the nonphysical essence of these bodies is acknowledged. This is a breakthrough in philosophical thinking.

This may be a breakthrough in philosophy, you may say, but it raises a lot of questions. For instance, why is mind-body duality a problem anyway? And why can't the mental and the supramental interact directly? And how does consciousness mediate nonlocally?

Well, let's deconstruct this and try to resolve some of these questions.

We have an external physical body that we experience in consensus with others and we have an internal mind that we experience privately. This is the basis of dualism—mental-physical, subtle-gross, mind-body, internal-external dualism; call it what you will. For millennia, it has been customary to assume that, since we experience these two things—mind and body—differently—one internally and the other externally—they must be made of different substances. In other words, mind exists as some kind of "subtle," nonmaterial substance. But this raises

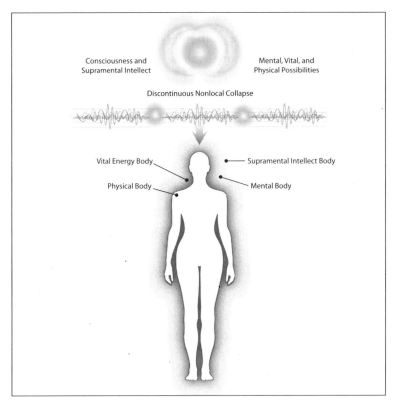

Figure 5. Pscho-physical parallelism; four experience types and four different worlds of experience.

the paradoxical question of how the nonmaterial and material interact, since they are assumed to have nothing in common. The solution to this paradox has always been that they need a mediator—some sort of signal—to interact. But materialist science claims that a signal carries energy, and that energy is a constant that neither leaves nor enters the physical world. This seems to rule out the idea of signals mediating any interaction between the nonmaterial mind and the material body.

Why can't the mind and the supramental interact? Because, although they are both subtle, they give different kinds of subtle experiences—thinking and intuiting respectively. Therefore, they, too, must be made of different kinds of subtle substance. So the question of how they can interact when they are assumed to have nothing in common is the same here as it was for mind-body interaction.

As for how consciousness can mediate nonlocally, this question is resolved in the quantum model, because all these worlds are possibilities of consciousness. Therefore, consciousness itself, as the common ground of them all, can mediate between them. And it can do so nonlocally—without signals—because they are all, in effect, a part of consciousness.

Materialists, of course, will maintain that the mind is the brain and that there is no difference between living and nonliving matter—between birds and rocks—on a molecular level. Therefore there is no need to postulate a nonmaterial mind or a vital body or the supramental realm. The relevant question is this: *What do the vital and the mental body do that the physical body cannot do?*

Vital Energy

Any sensitive person knows that, when we feel—as in an emotional thought—we feel energies. The spiritual traditions call this energy by a variety of names—*prana* in India, *chi* in China, *ki* in Japan, or simply vital energy in the West. We feel alive because we feel this vital energy. Some call vital energy the life force.

But feeling is not sensing. Sensing is the purview of the brain and the nervous system. Feeling occurs in conjunction with the organs of the body but is not actually of the body. Feelings are the movement of the vital body; the energy we *feel* is vital energy.

The concept of vital energy was discarded in Western biology and medicine because of its implied dualism and because, with the advent of molecular biology, it seemed that we could understand everything about life through the chemistry of DNA. But DNA alone cannot explain everything about the body—for instance, the many aspects of healing. As every physician and patient knows, healing often requires vitality—vital energy that is not the product of body chemistry. Chemistry is local, but the feelings of vital energy—the feeling of being alive—is definitely nonlocal. And where does vital energy come from if not from the movements of a nonmaterial vital body?

Molecules obey physical laws, but they know nothing about the contexts of living—maintenance and survival, or love and jealousy—that occupy us much of the time. The vital body belongs to a separate subtle world and contains the blueprints of form and function that define our fundamental vital functions—the contexts of living. In other words, the vital body provides these blueprints to the organs of the physical body that play out the vital functions of life in space and time. These blueprints are what biologist Rupert Sheldrake calls "morphogenetic fields."

The point I am making is this: physical objects obey causal laws, and that's all we need to know in order to analyze their behavior. We can call their behavior law-like. Biological systems obey the laws of physics, but they also perform certain purposive functions: self-reproduction, survival, maintenance of integrity of self vis-à-vis the environment, self-expression, evolution, and even self-knowledge. Some of these functions are instincts that we share with animals. For example, fear is a feeling that is connected with our survival instinct, but can you imagine a bundle of molecules being afraid? Molecular behavior can be explained completely within physical laws without applying the attribute of fear. Molecules do not cause fear. They are merely associated with the feeling of fear. Fear is a vital-body movement—something that we feel. When we feel fear in our vital bodies, a vital program is activated that helps consciousness guide the cells of a physical organ to carry out appropriate functions in response to a fear-producing stimulus, like the production of adrenalin.

Biological Blueprints

The behavior of biological organs is interesting because the blueprints—the programs—that run their functions are not related to the physical causal laws that govern the movement of their molecular substratum. Their behavior is thus program-like. Rupert Sheldrake's great contribution to biology was to recognize the source of this program-like behavior. He introduced nonlocal and nonphysical morphogenetic fields into biology to explain the programs that run biological morphogenesis—physical form and function for biological beings.

The point Sheldrake makes is this. We all begin as one-celled embryos that divide to make identical replicas with identical DNA and genes. But cellular functioning depends on the proteins the cells make. Potentially, all cells can make all the proteins, but they actually don't. Instead, cells become differentiated. Depending on the organ the cell belongs to, only certain genes are activated to make certain proteins that have to do with the functioning of that particular organ. So there must be programs, or blueprints, that activate the appropriate genes to produce the appropriate proteins.

How does each cell know where it is in the body and to which organ it belongs? The answer smacks of nonlocality. Sheldrake boldly suggested that the programs of cell differentiation needed for organ functioning require nonlocal (and hence nonphysical) morphogenetic fields. In other words, they communicate without signals.

The vital body is the reservoir of these morphogenetic fields, the blueprints of form and function. The job of the physical body is to make representations of the vital body's morphogenetic fields; these representations are the body's organs. The job of the representations is to perform the functions assigned to each organ—survival, maintenance, digestion, circulation, reproduction, etc. In this way, the vital blueprints provide the program for the genes that regulate the production of proteins suitable for carrying out the biological functions of the organ.

It makes sense. If living forms are run by software programs, then the programs must have started from blueprints developed by some programmer. The blueprints are now built into the hardware as form and function, and the program-like behavior of biological form is now automatic. So it is easy to forget the source of the program-like behavior and the programmer. And it's easy to forget that the behavior of biological beings is not always automatic. And it is also easy to denigrate the feelings that come from the movements of the source—the morphogenetic fields.

So the vital body is essential. It contains the original blueprints, the morphogenetic fields, that the physical body's organs represent. Once the representations are made, the blueprints are activated whenever the programs, which carry out the functions of their organ representations, run. The representation-maker, the programmer, is consciousness. Consciousness uses the vital blueprints to make physical representations of its vital functions that are codified in its supramental body, the body of laws and archetypes (see Figure 6). When consciousness collapses—or actualizes through choice—a physical organ to carry out a biological function, it also collapses, or actualizes, the vital blueprint. It is the movement of the vital blueprint that we feel as the vital energy of a feeling.

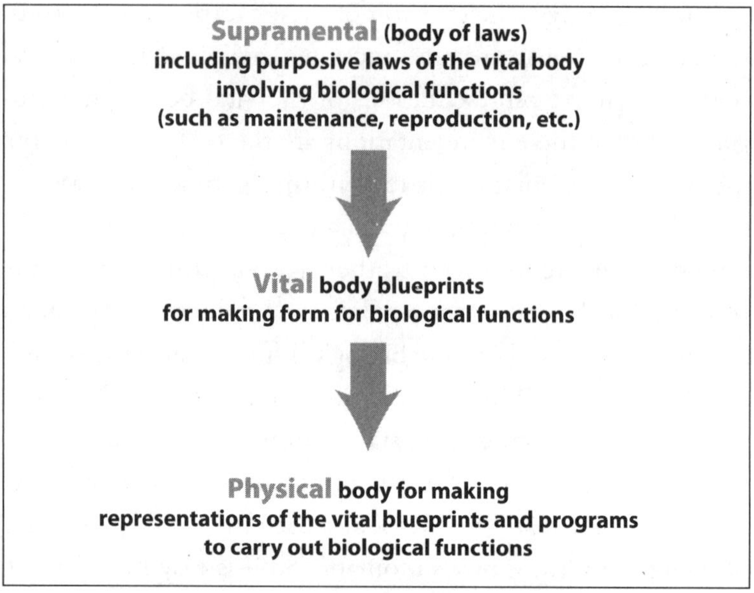

Supramental (body of laws)
including purposive laws of the vital body
involving biological functions
(such as maintenance, reproduction, etc.)

Vital body blueprints
for making form for biological functions

Physical body for making
representations of the vital blueprints and programs
to carry out biological functions

Figure 6. From the supramental to the vital to the physical.

Vital energy—or prana, or chi—is the quantum movement of the vital-body blueprint. When you experience an emotion internally, it involves thought, but it also involves an extra, subtle, vital movement that consciousness actualizes in your internal awareness. This is manifest prana. Emotions involve vital-body movements in addition to mental movements. Just watch yourself next time you are angry. Angry thoughts arise, But you also feel something else internally—something subtle. That's the prana, the vital energy.

Understanding the function and importance of the vital body gives us a profound explanation of feeling: what we feel, how we feel, and where we feel. But it is in alternative medicine that we find the most objective evidence for the importance of its role in our experience.

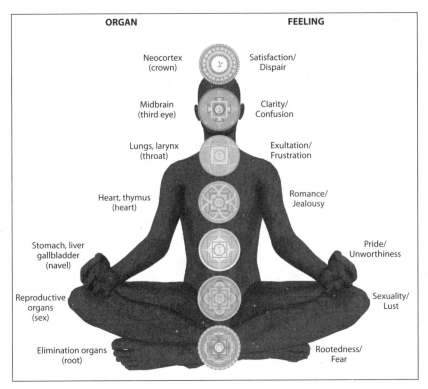

Figure 7. The seven major chakras.

One of the oldest alternative medicine traditions is based in a system of seven vital energy centers called the *chakras* (see Figure 7). Notice that each of these centers is located near a major organ and is associated with that organ's biological functioning. Each chakra is also associated with the feelings you may experience through the vital energy associated with that organ—the movements of its morphogenetic field. Each morphogenetic field is correlated with the organ of which it is the blueprint or source. Thus we are drawn to the conclusion that the chakras are the regions of the physical body where consciousness simultaneously collapses the movements of vital energy—the movements of important morphogenetic fields—along with the organs of the body that represent these energies.

So when it comes to emotions, the materialists have it all wrong. They think that we feel emotions in the brain—that is, that emotions are epiphenomena of the brain, the product of the combined action of instinctual circuits in the limbic brain and meaning circuits in the neocortex. Emotions, they claim, come to the body through the nervous system and the so-called "molecules of emotion." But actually, the body, in the form of the movements of the correlated morphogenetic fields, generates feelings quite independent of the brain. When we experience feelings at a chakra, the control goes to the brain for collapse and integration, because that is where the tangled hierarchy is. And the emotion is actualized in consciousness, because that is where the power of choice resides.

The Mind and the Brain

The neocortical part of the brain that is involved with mental phenomena like thought is a computer of sorts. So materialists ask if it is possible to build a computer that has a mind. This, they claim, would prove that the mind is simply a part of the physical brain—an epiphenomenon of the brain.

This misguided assumption spawned an entire field of study called artificial intelligence in the 1950s. Mathematician Alan Turing claimed that, if a computer could simulate a conversation intelligent enough to fool someone into thinking that he or she was talking to another human being, then we could not deny the computer's mental intelligence. Some claim to have accomplished just this. Moreover, one computer program has defeated one of the world's greatest chess players. But does that mean the computer is as, or even more, intelligent than a human being?

Enter philosopher John Searle. In a book aptly named *The Rediscovery of the Mind* (1994), Searle pointed out that a computer, as a symbol-processing machine, cannot process meaning from scratch. In other words, it cannot assign meaning to a symbol without a precedent. You can reserve certain symbols to denote meaning—call them meaning symbols. But then you need other symbols to tell you the meaning of the meaning symbols. So to process meaning starting from scratch, you need an infinite number of symbols and an infinite number of machines to process them. An impossible task!

Physicist and mathematician Roger Penrose gave a mathematical proof of Searle's thesis that computers cannot process meaning. In his 1991 book, *The Emperor's New Mind*, Penrose used Gödel's theorem—that any axiomatic mathematical system is either inconsistent or incomplete—to prove the cogency of the principle of tangled hierarchy. The theorem is a reminder that living matter, because it has to represent consciousness, has to be open-ended.

Materialist biologists claim that meaning may very well be an evolutionarily adaptive quality of matter. Searle's and Penrose's work convincingly expose the vacuous nature of such a claim. If matter cannot even process meaning, how can matter ever present a meaning-processing capacity for nature to select, whether it improved chances for survival or not?

The lesson from all this is that, although the mind is clearly associated with the brain, it does not belong *to* the brain nor is it caused *by* the brain. It is not an epiphenomenon of the brain. Instead, it is a body independent of the brain that gives meaning to our experiences. Computers cannot process meaning, but they can make (software) representations of the meaning we give them in certain contexts. Similarly, consciousness uses the brain to make representations of mental meaning.

Antagonists will argue that this is all theoretical and demand experimental data. And we do have a negative experimental test here. If this theory is incorrect, then it should be possible to build a computer that can process meaning from scratch. But while some computers can recognize cues given to them by a programmer, no one has yet been able to build a computer that processes meaning from scratch to refute the theory. In other words, the theory is passing the test. This may disappoint writers of recent science fiction who anticipate that robots will soon achieve consciousness and narrow the difference between themselves and humans. But that expectation only makes sense in a Newtonian worldview that ignores the principles of quantum science.

Meaning and Causation

The nature of brain memory clearly indicates that the mind is a separate entity different from the brain. Neurophysiologist Wilder Penfield first observed this while working with epileptic patients and stimulating their memory "engrams" with electrodes. He found that such stimulation produced an entire stream of mental memory. Mental meaning is thus represented in the brain, but only as triggers for the correlated mind to play back its correlated meaning. This also explains why memory is associational.

So what can the mind do that the brain can't do?

Let's start with creativity. A programmed brain can only handle what has been given to it—old meaning, not original meaning, or meaning from scratch. But creativity is the discovery or invention of *new* meaning. No brain, however it is programmed, can discover relativity or formulate quantum physics. Still, antagonists will argue that there is no causal practicality in the concept of meaning, so the question may be moot. But there are three important examples that show the causal practicality of meaning-processing: synchronicity, dreaming, and mind-body disease.

Synchronicity is a concept first introduced by Carl Jung. It refers to two events—one in the physical world and the other in the mental world—that are correlated through the meaning that arises in the mind. You can see an example of quantum nonlocality here. Synchronistic events are thus useful guideposts for the creative journey.

The neurophysiological explanation of dreams—that they are the result of putting perceptual images to white noise in the brain—is only the beginning of an explanation. The complete explanation is that the mind assigns meaning to the brain's white noise, sometimes creating quite striking audiovisuals. So dreams are the ongoing story of how meaning unfolds in our lives (see chapter 14). This explains why Jungian dream analysis, which assumes that every character in your dreams has a meaning that you give it, is so useful in psychotherapy. Because you assign the meaning, a dream can heal you when you work with it and appreciate that meaning.

There are creative dreams that have "disturbed the universe," like Niels Bohr's dream of the discrete orbits of atomic electrons in the form of an image like the solar system. Likewise, pharmacologist Otto Loewi got his inspiration for the experimental demonstration of the chemical mediation of nerve impulses from two dreams—first he dreamed the idea, but wrote it down illegibly; then he dreamed it again the next night and wrote it down more carefully. And there are more prosaic creative dreams as well. The inventor of the sewing machine, Elias Howe, got his crucial idea from a dream in which he was captured by savages carrying spears with holes near their sharp end.

Upon waking, Howe realized that the key to his machine was to use a needle with a hole near the point.

A third link between causation and meaning is found in the important field of mind-body disease. In somatic illness, errors in meaning processing can result in serious disease. (See my 2004 book, *The Quantum Doctor*. Physician Larry Dossey has written many books on this subject as well. See *Meaning and Medicine*, 1991.) For instance, cancer can result from immune system malfunction. Although there are always cells in our bodies that divide uncontrollably, with a healthy immune system, this is not a problem because the thymus gland makes sure that these abnormal cells are regularly killed off. So the suppression of emotion at the heart chakra, which is associated with the thymus gland, may contribute to cancer.

In the West, however, people, especially males, are culturally conditioned to suppress emotions. For example, a man may find that it is disadvantageous for him to have his heart chakra open in the presence of a woman that he likes, because an open heart makes him vulnerable. So he picks up the habit of suppressing vital energy at the heart, causing an energy block. A prolonged blockage like this impacts immune system activity so much that it can, in turn, suppress the body's ability to kill off abnormally growing cells that then become cancerous. Indeed, certain types of cancer have been connected with emotionally suppressive people blocking the energy of love at the heart chakra. There is now evidence that, when the emotions clear up—when a quantum leap in mental meaning unblocks vital energy at the appropriate chakra—patients can undergo spontaneous healing, taking a quantum leap from disease to wellness through their own creative choice. (See Deepak Chopra's *Quantum Healing*, 1990.)

If unbalanced processing of meaning can produce a serious disease in this way, and if right meaning can restore health, we better take mind and meaning seriously. They are not mere epiphenomena in search of a causal link!

Inner and Outer Space

If both mind and matter are quantum possibilities of consciousness, why is it that we experience matter as public (in outer space) and mind as private (in inner space)? Materialist scientists tell us that "the mind" is not a scientific concept, because we cannot study it objectively. Two people, they claim, cannot share the same thought and reach consensus about their mental experience. But what does quantum science say?

The materialists have no possible explanation for our inner experience, so they just wish it away as subjective epiphenomena needing no explanation. Nor do idealist philosophers, who value inner experience, offer us a cogent explanation. They just make the inner nature of the psyche a matter of metaphysical truth, and leave it at that. But in idealist philosophy, consciousness is the ground of being; all things are inside of consciousness—matter and psyche. So we are still left without an answer.

The quantum nature of the psyche, the mind, the vital body, and the supramental can give us an answer as to why we have inner experiences. Quantum objects are waves of possibility, expanding in potentiality whenever they are not collapsed. When I collapse a mental-meaning wave, I choose a particular meaning and a thought is born. But as soon as I stop thinking, the wave of possibility expands again. So between my thoughts and your thoughts, the wave of meaning expands to encompass so many possibilities that it is highly improbable that you and I will collapse, or actualize, the same thought. An exception occurs in mental telepathy, as we have seen. Another exception sometimes occurs when two people of similar conditioning converse. Generally speaking, however, thoughts are experienced as private or internal.

But why don't material objects, which are also quantum, behave in the same way? Shouldn't they also be internal to consciousness? In fact, if consciousness is the ground of being, why does anything exist outside of it. Maybe the idealists are right. Yet, we do experience matter externally. What is the mystery?

This is a crucial point. There is a fundamental difference between the subtle bodies and the gross physical body. The subtle bodies—the vital, the mental, and the supramental—are all one thing. They are each indivisible. But, as Descartes recognized, matter is *res extensa*, body with extension. Thus matter can be subdivided. In the material realm, micro matter makes up conglomerates of macro matter.

So, although quantum physics rules both domains of matter, micro and macro, there is a spectacular difference that arises when we consider macro matter as massive conglomerates of micro matter. According to quantum mathematics, a massive macrobody's wave of possibility becomes very sluggish. Suppose you and your friend are both looking at a chair. You collapse the chair's possibility wave and you see the chair over there by the window. Soon after, your friend also looks at the same chair. Between your collapse and your friend's collapse, the chair's possibility wave no doubt expands, but only a little. Moreover, the molecules of the chair are bound together with cohesive forces; so the chair-ness of the chair remains even in the domain of possibility. The center of mass of the chair can move due to the expansion of the chair's wave of possibility, but the movement is minuscule. As a result, when your friend collapses the chair, its new position is only minutely different from where you observed it, imperceptible without the help of a laser instrument. Naturally, you both think you are looking at the chair in the same place. You have a shared experience, so the chair must be outside of both of you.

Since the macro material world is built from micro matter in this way, it creates the illusion of remaining public all the time, even when no one is looking. And this is good in spite of the misunderstanding it creates, because otherwise we could not use material things as reference points. If your physical body were always depicting the uncertainties of quantum movement, who would you be? You would be like the Cheshire cat in *Alice in Wonderland*, appearing and disappearing and making the people you interact with dizzy! Moreover, if the quantum nature of macro matter were not subdued, how could we use matter for making representations of the subtle? Imagine writing your thoughts on a white board with a marker, only to see the marks move away in subsequent collapse events. What would that do to your representation-making capacity?

As to whether everything should exist inside of consciousness, this applies only when we are talking about nonlocal consciousness. We only experience matter outside of us from the point of view of our individual consciousness as locally represented in the brain. In mystical experiences, matter seems to be one with consciousness—the oneness experience.

Ever since Descartes recast reality as an internal-external, mind-matter dualism, Western philosophy has been saddled with this distinction. But quantum physics allows us to see that, like the Newtonian fixity of macrophysical reality and the behavioral nature of the conditioned ego, the interior-exterior dichotomy is also nothing but an illusion that masks the role of consciousness as reality. As we penetrate the illusion, we extend science to our subjective, interior experiences.

It is about time. Today we take much pride in people's increased awareness of the importance of ecology. The word "ecology" comes from two Greek words—*oikos,* meaning "where we live," and *logos,* meaning "knowledge." Ecology thus means knowledge of our environment, where we live. But where, in fact, do we live? Don't we live as much in our inner subtle space as we do in our outer world? Sociologist Erne Ness points this out and implores us to follow "deep ecology"—to learn to live in harmony, not only with our outer environment, but also with our inner environment. In other words, although we must take care of our outer world, we must also take care of our inner psyche. We must transform our inner being. The quantum worldview invites us to participate in this kind of dual caretaking, and quantum activists follow up on it.

CHAPTER 4

Zen and Quantum Physics

Zen Buddhism contains many parallels to quantum physics in the way it introduces the basic ideas of spiritual duality—heaven and earth, transcendent and immanent. For example, in Zen, there are two domains of reality—the domain of emptiness and the domain of form. The wave-particle theory of quantum physics recognizes two similar domains—the domain of potentiality and the domain of actuality. Likewise, consciousness plays a role in both realms in quantum physics, as it does in Zen, as shown in this parable:

> Two monks are arguing. One says, "The flag is moving."
> The other says, "No, the wind is moving."
> A master, passing by, admonishes them both: "The flag is not moving; the wind is not moving. Your mind is moving."

As stories like this show, students of Zen are often puzzled by their spiritual domains in the same way that students of quantum physics are puzzled when first confronted with the separate domains of the quantum world. Indeed, physicist Niels Bohr once famously said:, "If you are not puzzled by quantum physics, you can't possibly understand it." It is the same in Zen, whose students achieve understanding through a creative awakening. The discipline of quantum physics is not just a bunch of information to be learned. It is a way of looking at the world in which we discover the deeper implications of a new paradigm—one that gives us an awakening into the nature of reality itself.

Simultaneous Opposites

In the Zen way of thinking, opposites can exist simultaneously; contradictory things can exist at the same time, as revealed in this story:

> A master is teaching two disciples; a third disciple is sitting a little farther away, listening. One disciple expresses his understanding of what the master is teaching. The master says, "Yes, you are right."
>
> The other disciple, in his turn, gives a completely different interpretation of the teaching. Again, the master says, "You are right."
>
> Both students go away satisfied. The third disciple confronts the master, saying, "Master, you are getting old. How can they both be right?"
>
> The master looks at him and says, "You, too, are right."

Quantum physics works in a similar way. For every proposition, the opposite can also be true, because we always have this opposition of concepts being forced upon us by the nature of objects. For example, one of the very first teachings of quantum physics is that a quantum object can be both a wave and a particle. But waves spread out; they can be at two or more different places at the same time. Particles, on the other hand, behave differently. They can only be in one place at a time, and they always travel in a single definite trajectory.

In our everyday lives, we often face similar contradictory options. We want to make a decision—to choose—but we cannot because we also want to keep all options open. Quantum theory let's us do just that. We can keep all options open in potentiality, while making a decision—a choice—to collapse one potential into actuality. In psychotherapy, the domain of potentiality that contains all options at the same time is called the unconscious. More and more psychotherapists are realizing the value of the unconscious in therapy, finding that, when people allow their unconscious mind to process their choices, they are more satisfied with the result.

The important thing to recognize is that quantum physics is built into the nature of reality. When quantum physics says an object is both a wave and a particle, this is not a teaching tool or a metaphorical statement. It is really the case. For a long time, this was not clearly understood. Quantum theory was taken simply as a way of describing reality to make it more easily understandable. But this is not the case. Quantum physics is, in fact, a new way of recognizing quantum objects that is leading to many breakthroughs in many fields.

For example, take the case of wave-particle duality. When we say that an object is both wave and particle, we are not saying that an object is both wave and particle in space and time, in this very manifest space-time domain of reality. Instead, we are saying that the wave-ness of an object is true in a domain of reality that is beyond space and time—a reality that is unmanifest. We are saying that there is a domain of reality beyond space and time, a domain we call the domain of potentiality. In the domain of potentiality, the object is a wave of potentiality, or possibility.

The Domain of Potentiality

This domain beyond time and space is not only similar to consciousness; it *is* consciousness. This conceptual breakthrough came soon after experimental data showed that there is a way to distinguish, experimentally, between the domain of potentiality, where objects are waves of possibility, and the domain of actuality, in which objects are particles. French physicist Alan Aspect and his collaborators created an experiment that proved that this domain of potentiality has a unique, defining characteristic—communications that occur there do not require any signals, any mediation. The implications of this finding are astounding. For if communication can occur instantaneously and unmediated in this domain, it follows that the domain itself is *just one thing*. It is a continuum of interconnected stuff.

Here, we are communicating through words; I have written some words, and you are reading them using signals in space and time. But we could also be communicating through the domain of potentiality. If I formulate a thought but do not express it verbally or in writing, that thought can still be spread through the domain of potentiality and reach you. Instantly. This is what happens when we are inspired by an author's words or an artist's images to create a new thought or feeling. The written or spoken words, or the images, act as a trigger for a nonlocal connection that results in something entirely different.

In Zen Buddhism, we find riddles like this: What is the sound of one hand clapping? This riddle encapsulates the idea that things are born out of potentiality. Any thought is a potentiality with many meanings before it becomes an actual thought with one unique meaning. And in that potentiality, the possibility wave of that thought has many facets. The conversion of potentiality into actuality converts a many-faceted thought or object into a one-faceted thought or object—converts a wave into a particle.

Most of us tend to think that consciousness exists because there are human beings. According to the quantum explanation, however, consciousness already exists in the domain of potentiality, whether human beings are there or not. In fact, that is the whole point. But remember, this domain manifests. So we find that the manifestation of consciousness as self-awareness happens at the same time that thoughts or objects convert from waves into particles.

In this domain of potentiality, there is no form. Form is manifested in a specific way when a possibility is chosen and collapsed into actuality—manifest reality. So if we knew how to manifest a specific form in a specific way in the space-time domain—as a three-dimensional reality—we would be able to solve problems and actualize what we want in this reality. But this would require that we be able to sense or feel whatever the correct possibility is in the domain of potentiality.

And sometimes all we have is a feeling. We may have an intuition about what it is in the domain of potentiality that we want to manifest, but the domain of potentiality has many possibilities. So we have an opportunity to process all these various possibilities and their combinations simultaneously—an entire gestalt—in order to get an answer to the problem at hand.

This is where Zen and quantum physics converge as an approach to the human mind. Both Zen thinking and quantum thinking are based on allowing two levels of thought. By contrast, thought in a Newtonian world occurs on only one level. In this one-level world, which exists only in manifest space and time, there is only what we call conscious thought. Conscious thought lets us look at various possible answers, but we can only consider one aspect at a time, one facet at a time. When we allow the processing of thoughts to occur, not only in the space-time domain, but also in the domain of potentiality, *convergent* thinking can process many facets at the same time. The space-time domain is good for generating a whole array of divergent answers; we call it *divergent* thinking. But to arrive at a solution, simultaneous processing of many possibilities in the domain of potentiality followed by choice—convergent thinking—is more effective.

Now, the processing of thoughts is very different in the domain of potentiality. In space-time, we are conscious; in the domain of potentiality, we are unconscious. Only after repeated bouts of unconscious processing does convergent thinking manifest in the form of a solution—a quantum leap.

Multiple Possibilities

If quantum potentiality can contain multiple possibilities for what we are seeking, it follows that some of these possibilities will be "good" possibilities, and some will not. And of course, we always want to choose "good" possibilities—possibilities that will make things better or change our reality in a constructive way. So how can we be sure to retrieve the particular possibilities that will make for positive change amid all the potentiality?

This is a very good question, although the answer is not very satisfying: There simply is no guarantee. So our creative insights into a problem may have very painful consequences for others. No Japanese person needs to be reminded of the pain that the atomic bomb brought into the world. And yet the scientists who developed it certainly used quantum principles and certainly used Zen thinking. Sometimes, in the immediate aftermath of an event, it may seem as if creativity can lead to evil as well as good. But when we consider evolution—and we will—we see that temporary evil is sometimes necessary to get to the eventual good—to eventual progress through evolution. As painful as it may be for any Japanese person to remember Hiroshima and Nagasaki, the incidents did show us the horrors of atomic war and may have saved us from an even more devastating atomic war in the future.

And this applies to our personal lives as well. For instance, we enter into situations that may be challenging or difficult. But these are often the times when we can break through and reach the next stage in our personal growth. I am reminded here of another Zen story:

> A Zen master had the habit of holding up his forefinger, a habit that was often mocked by a certain little boy. One day, the boy held up his forefinger in imitation, and the master witnessed the act. He seized the boy and, using a sharp knife, chopped off the offending finger. When the boy cried out in agony, the master called out the boy's name to get his attention and held up his forefinger. The story goes that the boy attained enlightenment in that moment.

This story used to bother me a lot. It took me a long time to realize that its moral was that the boy just needed a jolt to move him to a new level in his personal growth. What seems like an evil is sometimes necessary to shake us out of ignorance and put us on the path to growth. Sometimes, unless we suffer, we never make that quantum leap into a better reality. In the same way, a dreadful disease can also be an opportunity to experience quantum healing—to make a quantum leap in emotional thought to heal ourselves by correcting faulty meaning processing (of feelings) and reactivating the immune system in full health. If we are ready, the same quantum leap can be a leap into enlightenment.

In the Zen tradition, students go through five days of very strenuous meditation, which causes a lot of pain in the knees. This pain can make students temporarily lose focus and let their minds wander. But once they have done the practice a few times, they learn to relax when the pain comes. The result is an alternating practice of doing and being—focused attention and relaxation—what I call do-be-do-be-do, after the Frank Sinatra lyric. After these five days, the students go into the presence of the master and experience *satori*, a quantum leap. Another Zen story considers this transformation:

> A group of Buddhists shared a spiritual practice of running around a mountain for one thousand days. But there was a certain monk who gave up after only a few hundred days. The others concluded that he must have been inspired and received a realization on the spot.

Creativity

The Zen path to sudden enlightenment was shrouded in mystery for a long time. But the mystery is resolved when we consider the quantum process of creativity. According to most people, creativity is work. Many people, and most scientists, think that all creative ideas are discovered because of the clever use of the so-called scientific method: *Try it and see.* Moreover, they try to glamorize the idea by claiming that, because scientists *try* and *see*, the process of verification is the crucial approach to creativity. But researchers have found that this is a very inefficient way of finding answers to really difficult and ambiguous questions, because there are just too many possibilities to try them all and to evaluate them individually. There has to be a better way.

After much research into many case histories, researchers found that something different was really going on. Scientists would work very hard to explore a problem. They would find some answers—some existing hints. And then they would just relax. Just relax. Do nothing. And often their breakthrough ideas sprang from that relaxed state.

A Japanese friend who was a copywriter in a high-powered advertising firm described some of his colleagues as more or less laid back, easygoing people. But he discovered that those seemingly easygoing, laid-back people seemed to be able to explore creativity much better than his busy, uptight colleagues. And that's the message of field research into creativity. Creativity not only requires the focus of driving intensity, which most people do have today but also requires a relaxed being—an unfocused mind. These are the two stages of the creative process: preparation and unconscious processing. Do-be-do-be-do.

Although researchers have known this for quite some time, no one could explain why it was so. When quantum physics and its proper interpretation came along, we found the explanation very easily. Between events of choice, quantum objects spread, because they are waves of possibility. Just as when you throw a pebble in the water, quantum waves of possibility literally expand and become bigger and bigger pools of possibility from which you can choose. Thus there is an advantage to waiting before choosing, because, if you choose quickly, the pool of possibilities from which you choose will be small. But if you wait—if you relax—that pool of possibilities will be considerably larger, and that is obviously a tremendous advantage for your creativity.

Of course, if you wait too long, you may simply lose your focus on the problem at hand. So there also has to be some sense of urgency that takes us to creative insights. That is why, when I teach courses that require term papers, I give my students deadlines. And I always encourage them to prepare and relax, but never to write the final version until the deadline is near.

For Better or Worse

In potentiality, there are many possibilities. Some of these possibilities will make things get better; some of them will make things get worse. Of course, we have a tendency only to look for the possibilities that will make things better; we tend to hope that only positive things will happen. In reality, however, it just doesn't happen that way. Negative things still do happen, in spite of our positive intentions. Perhaps we are inept at making choices; perhaps our intentions are confused. Will this change in the future, as we evolve in consciousness? Will negative things happen less frequently, and positive things happen more frequently as we evolve? My personal feeling is that, although on average that may be true, we will probably still occasionally need to actualize something negative in order to arrive at something positive later on.

The negative creates an urgency that generates intensity, which, in turn, spurs our creativity. Suffering creates motivation. Remember the boy who lost his finger. Of course, it would be nice if we could stimulate creative exploration through healthy curiosity alone. Unfortunately, however, in our present state of evolution, it is unlikely many of us will be motivated toward quantum creativity or Zen enlightenment by just honest curiosity, without some push forward from suffering. This is why I think, in Buddhism especially, there is emphasis on the recognition of suffering. And life and the world being what they are, I suspect that there will always be trials and tribulations that will spur us on to growth.

It is important to recognize that we can respond to crisis situations and negative events in a way that can promote more positive outcomes. Armed conflicts, global climate change, and political and economic problems can all be looked upon as danger signs that can prompt opportunities to create a new reality. We can look at a world full of problems and see it as an opportunity to make a creative quantum leap forward to a new worldview. In a worldview based on quantum principles and Zen wisdom, all things are possible—literally possible. From such a platform, we can make quantum leaps—discontinuous leaps of thought from potentiality to manifest actualities that have never been manifested before—to solve problems in physical and mental healthcare, in business, in politics, and in the environment.

Women seem to be better at this kind of leap than men. They are open to higher emotions; they are in their *hearts*. Likewise some cultures—like those of North India, Japan, Brazil, or Italy—are, in general, less logical and more emotional. This doesn't mean that they don't think logically, only that they tend to de-emphasize logic and think in a way that is more or less formless, more akin to potentiality. These cultures are ripe for quantum activism. When we mix the head and the heart, reason and emotions, we transcend both. With quantum creativity, we can resolve all kinds of conflicts and pave the way for a better reality.

As we try to make a quantum leap into this new reality, we will need both logical and nonlogical elements, emotions and intuitions. We will need to focus on feeling and intuition as well as on the mental and the logical. Zen and yoga flourished in America just as these practices were waning in Japan and India, and these disciplines present us with a way of thinking in which we can create a cloud of unknowing before we take the quantum leap to wisdom. Acknowledging this cloud of unknowing is an endeavor that will require intuitions and emotions to guide us. But we will also have to focus; we will have to bring conscious rational effort into the process of creativity. It will not be an easy process, nor will it be a quick one.

Let me close with yet another Zen story:

> A student had just finished fourteen years of awareness training. One rainy day, his Zen master invited him to his house for a celebratory dinner. As the student arrived, he put down his umbrella, took off his shoes, and went inside. The master welcomed him and asked, "Did you bring an umbrella?"
>
> "Yes, Master, I did," said the student.
>
> "You also took off your shoes, I see. Very thoughtful."
>
> "Thank you, Master."
>
> "Now tell me, on which side of the umbrella did you leave your shoes—the left or the right?"
>
> The student could not remember. "Master, I am not aware," he said.
>
> "Well, fourteen more years of training for you," said the master.

CHAPTER 5

Thought, Feeling, and Intuition

In the inanimate material world, the micro makes up the macro, and the macro is thus reducible to the micro. This "reductionism" is the way the material world is structured. In other worlds, in the worlds that we experience internally, there is no micro-macro distinction, no micro making up the macro. And that tells us something important.

One thing that modern science has done successfully is to show that the inanimate material world is reductionist—that all inanimate objects are built up from smaller components. And this has important benefits for us. The representation-making ability of matter is grounded in the fact that, at the macro level, there is hardly any quantum movement—there is approximate fixity. Without that approximate Newtonian fixity, we could not make representations of our subtle experiences using matter. And without representations, we could not have embodied consciousness (a tangled-hierarchical self) or the embodied subtle levels of experience (memory, software). In turn, without that embodiment and without memory, we would have no stability of personality, no written records, no civilization. Thus the micro-macro nature of material reality is really necessary for our world to function in the way that it does.

On the other hand, the exploration of the spiritual begins with the subtle. Spiritual traditions realized a long time ago that we can arrive at wholeness only through the subtle, not from the gross, material space-time reality. When we go to the subtle, we find that there is no distinction between micro and macro, no micro making up macro. If you get lost in the physical, you get lost in the simple hierarchies that matter creates, the hierarchies that make up the inanimate world. But wisdom—which has no hierarchies—is essential to attaining the spiritual domain of reality.

The wisdom of quantum thinking is that both the physical and the subtle are important; we need them both. Moreover, the four-fold nature of experience in the quantum model is more or less in tune with the four-fold Jungian classification of personality types—sensing, feeling, thinking, and intuiting. Jung showed that people selectively use one of these modes of experience in preference to the others. We need to change that tendency. We all need to integrate all these modes—to use all our experiences. Jung's work is important, however, because it codifies the way we are constructed, how we can experience. It takes the neocortex to collapse a quantum possibility into a subjective experience that consists, not only of sensing and feeling, but also of thinking and intuition.

The job of the neocortex is also to enable the mind to give meaning to all experiences. Unless we become very sensitive, however, what we really experience directly is not feeling, but emotion—which is the effect of feeling on the mind, feeling mixed with thought. We have to become sensitive to the pure feelings that preceded our experience of emotion. We have the same difficulty with intuition. We have to become sensitive to what actually happened before we can think about it intuitively.

Emotional Circuits

What actually happens when an intuitive thought comes to you? There's something very special about intuitive thought. It almost always occurs with a feeling in your gut or a feeling in your spine like a shiver. Or perhaps you feel a shaking in your knees, the location of a minor chakra. When we are sensitive, we can trace intuition back to its origin, to the experience that started both the intuitive thought and the gut feeling. Similarly, when we become sensitive to feeling, we follow the emotion back to its origin and begin to feel the vital energy viscerally in the body, at the chakras.

In general, creativity and intuition will come to you when you are more or less enjoying life and having fun. Intuition seldom comes to you in the midst of suffering or negativity, or in the middle of intellectual exploring. This is because suffering and negativity separate us from the unity that is higher consciousness, as does excessive rationality. In creative exploration, we are most successful when both the vital (the higher chakras) and the mental are engaged. That is why it is important to remember that what we call the supramental is also supravital. It also means that you are exploring fundamental creativity (creative exploration of the archetypes) and not situational creativity (solving a problem within given archetypal contexts). Situational creativity does not necessarily involve feelings. But to be most effective, we must involve the feeling dimension in all our explorations of creativity.

A philosopher friend once told me that he was able to learn to conjure up his feelings in a way that increased his efficiency greatly. On the other hand, sometimes using his emotions backfired, and he would be caught up in his own emotions and become anarchical—or even worse, *adversarial*. From this he concluded that, when you use it well, emotion creates wonderful results. When misused, however, emotion can lead to terrible consequences. Emotion is thus a two-edged sword. It can drive us in both positive and negative directions.

In short, it is easy to be confused about emotions. And this is a very important dimension to acknowledge. It is important to understand that feelings are connected to body organs. The picture of the chakras given in Figure 7 shows that there are seven major chakras along the spine—seven major centers of feelings. Each is located around very important organs of the body, organs that perform important biological functions.

Why is all this important? Remember the morphogenetic fields? They are blueprints of biological form and function that help consciousness create the organs that carry out biological functions as life evolves. Moreover, the feeling that you experience at each chakra is strongly correlated with the biological functions of the organs at each location.

But it is an important part of evolution that, when the brain—whose function is to integrate—evolved, it took over some of the control of the functions of the body. The price we paid for part of that integration, the coordination process, is that the brain developed as the center, not only of thoughts, but also of feelings—at least those feelings connected with the lower chakras. These evolution-conditioned feelings were represented in the brain.

As the mind gave meaning to feelings, they became associated with neocortical thoughts. The negativity of an emotion comes from the association of feeling with the mind—mental meaning. We have what are called instinctual brain circuits that yield negative emotions like anger, lust, competitiveness, jealousy, or envy. But evolution did not yield many positive emotional brain circuits—just a very few, like the maternal instinct. In fact, in most cultures, we find that the maternal instinct is worshipped as a sacred reflection of the archetype of the divine mother. Some people also have what can be called an altruistic brain circuit. And there is also a spiritual brain circuit—hailed as "God in the brain" by some scientists—which, when excited, triggers a "spiritual" experience. However, there is no brain circuit for unconditional love, no brain circuit for goodness in general, or for beauty, or abundance, or justice, or wholeness.

As long as we remain insensitive to what is happening in our bodies when we are in the throws of an emotion, we will remain unaware of our pure feelings. And that presents a difficulty for most of us. By pure feeling, I mean a feeling without any meaning given to it by the mind—feeling without thought. And this is the very important distinction between emotion and feeling. Emotion is feeling plus thought, feeling plus the meaning given to it. This is is what Jung was getting at when he established that there are four different kinds of pure experience.

Chakra Energy

One of our first tasks as quantum activists is to become aware of the visceral aspects of emotions—pure feelings we experience when an emotion arises. We can do exercises to activate the chakras and become familiar with them. With practice, we can become sensitive enough to feel the energy of the chakra preceding the brain experience. It is not easy to feel the energy at the lower chakras when you are in the throws of an emotion. But the advantage of doing so is that, once you get the hang of it, you can also feel the energy in the higher chakras—feelings for which there are no instinctual brain circuits.

In this way, you discover something very beautiful—that the brain does not easily dominate the higher chakras. We must learn to activate these chakras. For example, we can do exercises like picking up a baby or a pet and feeling the energy in our hearts. When we do this, the tangled-hierarchical brain picks up the feeling, consciousness gives meaning through the mind, which the brain records, and we experience the positive emotion of love. And eventually of course, this experience will be lodged in a brain circuit of memory.

Or you can practice singing in the bathtub. You feel the energy in your throat chakra, which in turn generates your eventual experience of the positive emotion of exulted expression. You may feel as though winds are blowing in your throat chakra. Your throat may tingle, or you may experience a stronger sort of movement, like a throb. The tiles in a shower or bathtub make the sound reverberate and amplify it, making it stronger. If you sing without reservation, you start feeling energy in your throat chakra. If you sing a love song, you can activate your heart chakra. The brow and crown chakras are a little more difficult to activate, but you can use the heart and throat chakras to help activate them and let them lead you to a world of positive emotions.

At the higher chakras, consciousness will seem more expanded. You will feel warmth that comes from excess vital energy in those chakras. For example, when clarity comes to us, we feel warmth between our brows. This warmth can be quite intense. At puberty, some girls begin to intuit their personal power, for which the vital energy rises in the brow chakra. In India, the custom is for girls to wear some covering, like sandalwood paste, over the brow chakra because the covering is thought to keep the place cool. And even now, Indian women wear very stylish coverings called *bindis* between their eyebrows.

Years ago, my wife and I taught joint workshops on quantum physics and yoga psychology in Scandinavia. She used to wear exotic bindis to the sessions. Our clients, at least the women among them, must have been very curious, but they never asked about the bindis. One day, as I was explaining the original use of bindis, my wife entered the classroom to teach the next session wearing the most exotic bindi you ever saw. And the women clapped. They understood.

Most exercises for activating the chakras are designed, not so much for the feelings generated by the lower chakras, but for focusing more on higher chakras. The lower chakras are not considered as useful and can even be considered hazardous. But they are also helpful to us in some ways. We can start managing the lower chakras as part of our health regimen. A huge problem of our body-mind system is that we experience our emotions exclusively in the brain, so we often neglect our bodies. And because people neglect their bodies, diseases caused by the malfunction of the lower chakras are very common. For example, many people today suffer from irritable bowel syndrome or constipation that Freudians ascribe to anal retentivity from bottled-up emotions. These are lower-chakra diseases that occur because we are not balancing these chakras, and thus not ensuring a free flow of vital energy through them.

Similarly, many people, especially women, have difficulties with the third chakra—the stomach chakra—that arise from insecurity. Ulcers, for instance, are a very common disease that comes from a third-chakra energy blockage. These diseases are often exacerbated because we don't pay attention to experiencing the energy at these chakras in the body. What we experience is the emotion associated with a brain circuit—anger, pride, or an overdose of egotism. All this we experience in the brain and not in the gut.

So what are some safe ways to experience activated feelings generated from the lower three chakras? First, you must be sensitive to the body as you emote, beginning with the brain. You experience emotion first, no question, because the brain takes over—your sense of self is located there. But each emotion comes with a history. Something occurred before the emotion. When you become sensitive, you remember the entire chain of events—events that consciousness chose and collapsed—going backward in time. Emotions first start with feelings in the lower part of the brain stem—the cerebellum, pons, and medulla oblongata. Eventually, the neocortex becomes engaged.

Choice and subsequent collapse—the change of possibility into actuality—can precipitate an immediate isolated event. But it can also precipitate an event retroactively, by going backward in time and following an entire chain of events that are prerequisite to the present event. This is called *delayed choice*, or *delayed collapse*. Consciousness chooses with a delay, collapsing an entire chain of potentialities and producing an entire chain of events going backward in time, all the way to the potentiality that started the causal chain. This idea of delayed choice is a completely logical idea that has been verified by experiments that I talk about later. Here, suffice it to say that, when we look at the memory prior to an intuitive thought, we can become sensitive to intuition and to the feelings that arise in the body when we experience an intuition. When we do that, we open the way for experiencing the higher-chakra feelings.

Take love, for example. What is the feeling that most often obstructs love? What obstructs it most is the feeling of fear. Fear can prevent us from dissolving into love. Or sometimes it is ego-defensiveness. When confronted with the emotion of love, we think: "I want to defend who I am" or "Why should

I make a difference for somebody else?" Which, by the way, is exactly what love wants us to do. So to reenforce our feelings of ego and control, we choose sexuality rather than an exploration of archetypal love.

We cannot avoid negative emotions like anger and associated attack words. Those negative emotions seem to happen automatically. But what we can do is to pay attention when negative emotions happen. We can pay attention to our vital-energy feelings at the lower chakras—which may seem like physical sensations but are not—and their associated feelings. But then what do we do? Pay attention to our lower chakras? Focus on the heart? Try to raise the emotion to an even higher vibration—to the higher chakras?

I think the best strategy is to approach this in several steps. In the first step, we can develop body sensitivity. Once we have that, we can consciously create positive body experiences of feeling with the help of the mind and memories, or through direct experience—like holding a baby and feeling the energy that comes.

In the second step, we can learn simple things—like focusing on a problem and becoming aware that the third-eye chakra is involved. When we experience clarity, we focus. We become aware that we are squeezing the muscles between our brows, or that a central place in the forehead is heating up. We can even do something to activate the crown chakra. We can engage in activities that satisfy us, feeling the energy viscerally in the brain when we are being satisfied.

In the third step, we can begin to use the vital energy exercises of certain sophisticated traditions like Chinese *tai chi* and *Chi Gong*, Japanese *aikido* and martial arts, and Indian *pranayama*. Pranayama uses breath control to carry air all the way to the brow chakra and then down all the way to the navel chakra.

This employs a technique called *deep inhalation*, in which we inhale as if the air is going all the way to the top of the nose while visualizing that the brow chakra is activated with that flow of air. And when we breathe out, we empty the stomach, squeezing the muscles there, and feel the energy in the navel chakra. Likewise, tai chi, Chi Gong, and aikido use gestures and arm-and-hand movements to raise the energy from the root chakra all the way to the crown chakra. There are simple exercises for beginners in these traditions—techniques that I teach in my workshops to acquaint people with vital energy and how to vitalize the chakras.

I will end this chapter with a personal anecdote of how I discovered the movement of energy in my own chakras. In 1981, I was guest lecturer at the Esalen Institute in Big Sur, California, for a week. The late spiritual teacher Osho (then Bhagwan Shri Rajneesh) had a big following in America at the time. I was invited for a morning meditation with a Rajneesh group and I went. Someone explained that the meditation would consist of four parts. We began with shaking our bodies standing in place; I discovered this really wakes you up. In the second stage, we were told to stop right where we were and meditate in that position. In the third stage, we began slow dancing to music with our eyes closed. I was doing fine, until I bumped into someone. I opened my eyes and was confronted by a bouncing pair of breasts! (The Esalen Institute was quite famous then for allowing a lot of nudity, and, in spite of being in the United States for some time, I was still unused to it.) Well, my body reacted by producing that peculiar protrusion of which a man's body is capable, and I was thoroughly embarrassed. Fortunately, the bell rang to signal the beginning of the fourth stage, in which we all sat down and meditated. But my feeling of embarrassment persisted and generated a strong feeling of energy that rose from my anus to around my throat. And it was quite delightful.

Now mind you, I grew up in India and, in that culture, everybody knows about kundalini awakening, in which you supposedly experience the movement of prana (vital energy) from the lowest chakra to the highest. The Sanskrit word *kundalini*, in fact, means "coiled up energy." The same idea, expressed in quantum terms, is that the feelings at these chakras remain in potentiality until a sudden quantum leap of awakening takes place. So at once I suspected that I had had a kundalini awakening experience. But I was disappointed that the energy did not rise all the way to my crown chakra, as the literature describes. Was this a kundalini awakening? I do not know. But it certainly made me aware of the chakras; from then on, I knew their energies were real.

CHAPTER 6

The World of Archetypes

Beyond the mind is the world of archetypes that philosopher Sri Aurobindo called the "supramental." Actually, this is an incomplete description of that world, because the realm of experience beyond mind also goes beyond the vital energies that we feel. We can represent these archetypes in feelings as well—what we call our noble feelings. So supramental is also supravital, and it provides really interesting contexts for our mental thinking and our vital feeling. We should really call this realm the supramental/vital world.

Intuitions are a glimpse at this supramental world—the archetypal world of love, beauty, truth, goodness, justice, wholeness, abundance, and self. Unless we learn to be sensitive to the subtlety of intuitions, we cannot really explore this supramental world. When we become sensitive and follow through, however, we engage in the creative process. When we follow through on our intuitions all the way to the essence of an archetype, we make our own mental representation of our insight into the archetype and eventually develop it into a product that others can appreciate. I call this "fundamental creativity." More often, however, people creatively interpret archetypes from the experiences of others—reading a book, for example. Upon further exploration, they receive a secondary insight into the archetype in the context given by that vicarious experience. This is what I call "situational creativity." In situational creativity, we try to solve a problem by staying at the same level in which the problem was created.

Archetypes and Intuition

The quantum worldview assumes that all the worlds of quantum possibilities from which our experiences flow are located within consciousness in the form of the unconscious. We have four worlds of potentiality from which our four different kinds of manifest experiences arise (see Figure 5). Of course, the physical world is the one of which we are most keenly aware. Our sensory experiences come from our physical responses to potentialities that we call physical stimuli. But where do our feelings come from? They do not come from the physical world. They come from the movement of energies in the vital world of potentiality. Of course, the physical may influence the vital through the intermediary of consciousness. And the vital influences the physical through the intermediary of consciousness as well. So these potential worlds are parallel, interacting worlds. Thinking comes from the mental world of potentiality. When consciousness chooses from possibilities of meaning, we get a thought. In the potential world of archetypes, however, the objects of possibility are archetypes.

Can archetypes be objects of possibility—waves of possibility? They really are multifaceted objects. That is the reason we discover so many different facets of the same archetype in different instances. There is only one exception—the archetype of truth. Truth is absolute. The archetypal world is a world of truth.

Why is truth not multifaceted? Because the universe needs to be created with a fixed set of highly refined laws. So, all the objects of the archetypal domain, all archetypes, have "truth value." This is why, in creativity, in a creative experience, we always know what we know with *certainty*.

And when we, as nonlocal consciousness, choose a particular facet from these archetypal possibilities, we experience an intuition. But our physical bodies don't have any ability to immediately represent the intuition experience—that is, to make a direct memory of it. So we make a mental representation of the intuition. The brain is able to make a memory of that mental representation so that what we call a quantum measurement can occur. When we remember an intuitive experience, we are really remembering a mental representation of the intuition.

When we explore intuition, we are exploring the leading edge of evolution. From the point of view of consciousness, evolution is not an evolution of matter as Darwin saw it. Instead, it is the evolution of material representations of consciousness and its subtle potentialities (see Figure 8). The vital is represented first—the evolution of life from a single cell all the way to human beings. Then, with the advent of the neocortex, the mind can be represented. Before the advent of the neocortex, animals had minds, but they experienced the mind by making representations with their feelings. Similarly, at our stage of evolution, we represent archetypes with the mind and the vital energies, and then with the brain and the body's organs respectively.

When a supramental archetype visits us as an experience of intuition, it comes to us through the intermediary of the mind. The mind gives meaning to the archetypal experience and the brain, the neocortex, makes a representation of the mental meaning. Representation here means just memory—a memory of mental meaning. When that memory is triggered, the mind plays back the meaning. Eventually, we will evolve to a point where we can embody the archetypes directly.

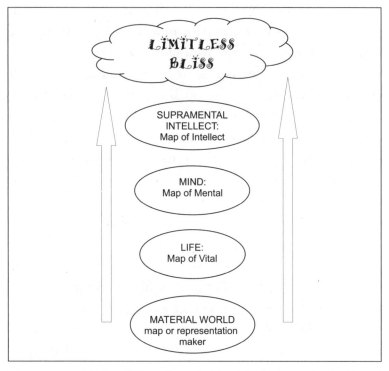

Figure 8. The evolution of life as the evolution of the representation-making of subtle potentialities. (Illustration by Terry Way)

Quantum Evolution

Let's talk more about the process of evolution through a quantum lens. Matter builds complex forms in a simple hierarchy—from elementary particles to atoms to molecules to bulk matter. The causal flow is one way, from the bottom up—and all this occurs in potentiality. Then molecules of matter make a tangled hierarchy of two components—neither bottom up nor top down, but both partners in a circular causation. When such a tangled hierarchy in possibility matches the vital blueprint of life, consciousness recognizes the match and life begins with conscious collapse.

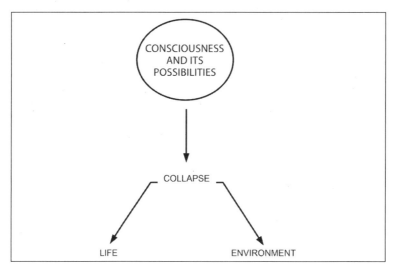

Figure 9. How quantum collapse creates life and its environment. (Illustration by Terry Way)

In this way, in a living cell, the macromolecules of DNA and protein are the molecules that make a tangled hierarchy of circular causality—DNA is needed to make proteins, but protein is required to make DNA. Circular causality seems to occur in a self-contained system. This is the basis of the self-making of life that biophilosopher Humberto Maturana calls "autopoiesis." Consciousness is thus represented in the cell in a rudimentary way; it has identified itself with the cell and experiences itself as separate from the environment—all the other molecules in its surroundings (see Figure 9).

Living beings initially make representations only of the vital. Although the cellular consciousness can feel, the self that feels is rather primitive. As evolution proceeds, better and better representations of the vital energies appear in form, and better and better organs are actualized to carry out biological functions. So we grow from a single cell to multiple cells to vertebrates to mammals to primates to humans. Somewhere along this journey, the evolution of the neocortex occurs. The neocortex is complex enough to create a thinking individual self much more sophisticated than the cellular self. In other words, it has a more sophisticated tangled hierarchy built into it.

This self has access, not only to vital feelings, but also to mental meaning, because mental meaning is represented in the neocortex. And as evolution proceeds, the mind can give meaning, not only to what were before the sensory stimuli and vital feelings, but now also to mind itself—meaning itself—and even to the supramental archetypes that will be represented in matter only at a later stage of evolution. And whenever the mind gives meaning, the brain makes a representation of that meaning, that particular thought (see Figure 10).

From Experience to Archetype

Now let's return to the psychophysical parallelism shown in Figure 5. At the top is consciousness—nonlocal consciousness. This is the domain of potentiality, which contains four kinds of possibilities: physical possibilities, vital possibilities, mental possibilities, and the supramental. Collapse from these four kinds of possibilities can generate four kinds of experiences: sensing, feeling, thinking, and intuiting.

As an example, let's assume that you meet someone to whom you are attracted. At first, you see the physical body, which is a sensory image. But you may also have a feeling in the lower chakras—a sexual feeling that you experience as an emotion. Simultaneously, an intuition may come: I would like to explore the archetype of love through this person. This is translated as a thought: I would like to get to know this person. Notice that the mind gives meaning to the sexual feeling. But it also represents the intuitive idea, which is quite different: I would like

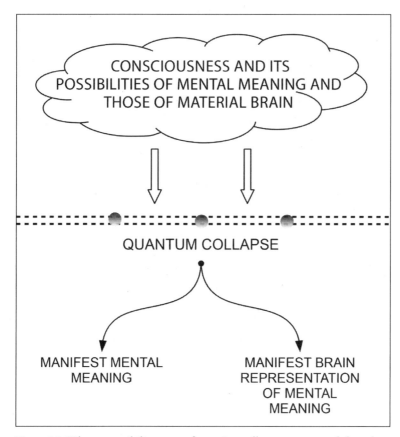

Figure 10. When a possibility wave of meaning collapses as a mental thought, the brain makes a representation of it. (Illustration by Terry Way)

to explore love with this person. Part of this idea is often left covert, but it still may be the beginning of an exploration of love, because the intuition generates an additional "gut" feeling that there is something here on which you should follow through. In this way, the same physical stimulus leads to all four kinds of experience. Consciousness mediates the interactions of the different potentialities whenever necessary. It also chooses and collapses each one of these experiences.

So intuition is the beginning of creativity. If you follow up the intuitive thought (I would like to know this person), it may involve creative exploration of the archetype of love. What is the first thing that may happen if you follow up that intuition? You may think: How can I get to know this person? Then immediately, you examine your feelings and find that there is a sexual feeling involved. That's just brain circuitry. But then you remember that there was something else—you had some feeling in your gut that revealed the involvement of your heart chakra. This feeling in the heart chakra is a delayed choice that preceded the experience of the gut feeling. But you were not sensitive to it—perhaps because of habit, perhaps because of fear.

Usually our experiences of the chakras are superseded by the experiences of the brain. But if we learn to be sensitive, we become aware that the heart chakra feels as if it were fluttering. Or perhaps tingling. Or perhaps expanded. And that's when we do a double take. We look again to confirm. And this time, the experience is quite different. This time we are sensitive to the heart. And we recognize that there definitely is something there. And then we are motivated to follow it up.

So how do we follow it up? The predominant feeling will usually be of bodily union—a bodily union made possible because the immune systems of the two people involved, in the form of their thymus glands, the most important heart-chakra organ, are cooperating. The "me-not me" distinction has been suspended. But there is a brain component as well. The sexual brain circuits play a huge role by creating lots of endorphin molecules and feelings of pleasure. Pleasure becomes a major motivation. Unfortunately, these pleasure molecules, these neurochemicals, run out sooner or later. For some, they run out pretty quickly; for some, they run out more gradually.

After the neurochemicals run out, you have a choice. And this is where, if you want to explore love, you have to make a decision to follow up. The choice to follow up means commitment.

If you choose commitment rather than moving on, you begin an exploration of unconditional love. This commitment is a very important part of creativity. You have to stay with the exploration. Creative quantum leaps cannot be manufactured by wishing them. They are discontinuous. Therefore nothing that you can do with continuity can guarantee that you will have a discontinuous thought or jump—a creative leap. This is why, in most cultures, this commitment leads to a social agreement called marriage. Today, in more and more cultures, even gay couples can commit to each other—legally through marriage, as well as romantically through their commitment to explore unconditional love. The evolutionary movement of consciousness has gotten rid of an age-old taboo of human societies to accelerate the exploration of love. So we consider marriage a partnership—a committed partnership. Within that committed framework, the creative process can begin in earnest, with focus.

But we have to be careful here. Because whether the archetypal context of a committed relationship becomes important can affect the degree of difficulty of our creative journey. In the case of another human being with whom you are exploring love, the context is clearly an archetypal context. But in the case of a business relationship—a client relationsip, for example—the archetypal context already is represented in the form of set, or even contractual, expectations. This kind of known context triggers what we call situational creativity. When you begin an exploration of love in a human relationship, however, situational creativity will not work because it limits the possibilities you can create about love. If you work within the social norms of marital behavior, very soon you will organize yourself into love within those rules. That is not creativity—or at best, it is situational creativity. The opportunity present in a committed relationship is that you can explore deeper levels of love without role playing, without societal expectations, without even the expectation of sexuality. So exploring love in human relationships should always be approached with fundamental creativity in mind.

But how should business relationships be explored, or problem solving, or product development? Through brainstorming? And if so, what decides whether or not an insight—a quantum leap—occurs? Remember that creativity, as we have studied it so far, is mental. It is the discovery of new meaning. And that new meaning may have a new business application. Insights always come to an individual, but the presence of a collective can create a stronger intention and better access to quantum consciousness. So brainstorming may be effective for making business decisions or gaining creative insights, provided that the people participating are in tune with one another.

But the archetypal contexts of human relationships cannot be explored by staying within old contexts and meanings. This requires fundamental creativity, which takes you beyond the level at which the situation was created. No one can tell you how to love unconditionally. All they can tell you is what works for them. And if you think about it, that will never work for you, because the conditions of your situation are very different. Therefore, you really need to look directly at the archetype yourself to learn what unconditional love entails in your particular context. And that is what I call the exploration of fundamental creativity.

Archetypal Contexts

By examining meaning in the context of its archetype, you also explore your consciousness. The exploration of an archetype entails quite a bit of exploration of your self. And this involves inner creativity. What you explore is the archetype—the subtlest level of consciousness that we know. What you create is not only the meaning, but also the archetypal context. You make a representation of the archetype in your being. In a sense, you find your own personal version of it—your own version that will dictate your own behavior. When you do this, you become an original.

Take love, for instance. Even in the exploration of creativity in general, we can make love a crucial part. *And it will be better for all your exploration.* We can seek to explore all archetypes, with love always present in the background. Many spiritual traditions do that, because love is such a fundamental archetype. Suppose we explore other archetypes—beauty or goodness or abundance. Creating abundance interests many people in our society. But can you imagine the impact of exploring abundance with love in the background? Then we would never seek abundance that would harm other people or be achieved at the expense of other people. We would automatically become sensitive to the ecosystem and to ethics. We would focus on a very important thing that is lacking today in our exploration of archetypes, especially in religion and politics.

Wisdom from *The Wizard of Id:* The wizard tells a friend he has been working on a way to kill antibiotic-resistant bacteria. The friend, surprised that it worked, asked how he did it. The wizard replies that he introduced them to religion and politics.

Suppose you are a college student trying to choose a career. You are looking for an occupation that will make you feel abundance—feel rich. How can you make the most money? But as a student, you also want to be creative—you want to become somebody. You also want to get some intuition and use creativity to decide what career you are going to explore. With all these materialistic expectations and goals, how can the archetype of love help you to achieve what you want?

Well, let's examine what is actually happening here. By seeking abundance on the material plane (money) alone, you have already limited yourself to situational creativity. So love is excluded from the beginning. If you seek abundance without the context of love, you may eventually choose to go to Wall Street and become an investment banker. But if you seek abundance with the background of unconditional love, you may found an ecological enterprise or some sustainable business or some company that is helpful to society.

This is the crucial point. If you choose money as a representation of abundance—if you choose information instead of meaning as the basis of your creative search—then investment banking may be the best profession for you. But if you make your choice in an archetypal context rather than in terms of the material representation of money, you will see that money is not the only form of abundance. You can also be abundant in love. You can find an abundance of meaning in your life. So the archetypal context of abundance changes the equation for you, and you will become attracted to archetypes like love, because abundance will include those energies.

There are very important differences between situations where we use scientific materialism to guide us and those where we use a much more general philosophy of consciousness to guide us. In scientific materialism, meaning is not encouraged. Instead, information is encouraged. The digital age has, in fact, given us the false impression that everything is just information. So, if someone tries to bring love into a career choice, the result will be representations of love from all over the Internet, which is just information—information without meaning. This is how the materialist pursuit short-changes not only the archetypal value, but also the meaning value. It emphasizes only the barest informational value. This is why innovation, the entire culture of business innovation that used to be evident all over the world, is now suffering.

But is love the only archetype that can activate creativity, or are there other forces that can activate it? Any intuition is an archetype calling us to explore it. So any archetype can trigger creativity. As we saw above, the archetype of abundance can naturally lead to the archetype of love. And it can also lead to the archetype of goodness. For an artist, the archetype of beauty may be the initial attraction, and that can lead to the archetype of love. This can result in art that represents not only beauty, but also love. And that is the kind of art that we all covet.

Love, Self, and Wholeness

Love is certainly the most ubiquitous and the most sought-after archetype. But is it the most fundamental and important one? There are actually three fundamental archetypes that contribute to our consciousness realm. The first is love. The second is self. The third is wholeness. Love is the most attractive one, and most people see the necessity of investigating it early on. But as we grow psychologically, we begin to see the importance of the self archetype. As we become more psychologically mature, we are confronted by the limitations of the self, or the ego, and begin to ask whether we can move beyond those limitations. Though we may not yet be interested in anything like enlightenment in the sense of the spiritual traditions, we begin to ask basic questions about human potentiality. We begin to explore the possibility of expanding our horizons.

At about the same time, we also become aware of the archetype of wholeness. When you lack wholeness, when you get sick, you may look at the illness as a sort of suffering—as an inconvenience. So you go to the doctor and ask to be cured. The doctor may give you some pills or attempt some other quick fix, and that's it. But at some point in your maturing process, you begin to recognize that curing is not the same as healing. Curing a disease does not necessarily make you well. At that point, you become interested in health as wholeness—health, not only as an absence of suffering, but also as an optimization of well-being. Then you are ready to explore the archetype of wholeness.

If love and self and wholeness are fundamental, crucial archetypes, why do they not play a greater part in our experience? In recent history, we seem to have missed the importance of these archetypes, first because of a religious worldview that falsely interpreted the archetype of wholeness. Religions tend to divide the world into the spiritual and the material—just the opposite of the materialist fallacy. They ignore our external well-being in favor of our internal well-being. Materialists choose a material representation of abundance; spiritualists choose a spiritual representation of abundance. Neither of them acknowledges wholeness.

The spiritual worldview has guided us for thousands of years and some good has come out of it. But four hundred years ago, when modern science came onto the scene, we made a holistic compromise. Under what is called the philosophy of modernism, we valued both matter and mind, spirituality and materiality, combining them into a dualistic worldview that kept mind and matter separate. That worldview gave us the institutions of democracy, capitalism, and liberal education, and had an enormous impact on developing human civilization. In fact, the quantum worldview is nothing but an integrated form of modernism. Some people call it transcendent modernism, or *transmodernism*. This worldview integrates the dualistic mind and matter of modernism within consciousness. Unfortunately, instead of going from modernism directly into the quantum worldview, we got side-tracked into scientific materialism—an unfortunate mistake for humanity.

The film *The Wolf of Wall Street* is based on the true story of an investment banker named Wolf who started from very poor conditions. Through some crude sting operations, he became quite rich. In the end, however, he was prosecuted by the FBI and jailed. After spending time in jail, he became a teacher of self-growth and healing. But even when he was doing lots of bad things, Wolf had radiance. He had the basic good instincts within him, and just needed suffering as a trigger to wake him up to the archetype of goodness.

When we create evil in our lives, it often provides the springboard for a very difficult transition about which we are confused. That's the beauty of life. Nothing in life can be said to be entirely meaningless. Life always has some meaning, because it always has the capacity to teach us something. If we are open, we can learn from the negative experiences that scientific materialism has brought us, and use them as the springboard for really satisfying explorations of the archetypes in a new way.

CHAPTER 7

The Ego and the Quantum Self

The self we ordinarily experience is called the ego. The consciousness that can get trapped in the tangled-hierarchical brain in the form of a self is not the ego, since it has no memory and conditioning, no personality. The ego personality, on the other hand, is the result of conditioning. But how does conditioning enter the quantum view of the self?

Experience of the ego self comes to us in two stages. The first is the result of tangled-hierarchical collapse, as I explained before. This is what we call the quantum self. If only one collapse, or choice, were to occur, this would be the extent of the self.

But every time a stimulus comes and a new response occurs, the brain makes a memory of the event. And every time a stimulus is repeated, consciousness responds, not only to the primary stimulus (the quantum self's response), but also to secondary responses that are stored in memory. This cumulative reflection of events in the mirror of memory is what produces conditioning. The self of this conditioned response is the ego.

Before conditioning occurs, the world-experiencing subject, or self, is unitive—one subject for all. You can also call it cosmic. Experience at this level of self is a very special experience—you feel one with everything.

But we see new things all the time when we go to new places. Yet we may not experience anything special about seeing a new waterfall or a new river. This is because we have seen waterfalls and rivers before and our minds have become jaded. If someone has never seen a waterfall, the experience can be spectacular. But even then, some people may miss the immediacy of the experience, because their minds get busy comparing the waterfall with similar past experiences, or with what they imagined the experience would be like.

This is why poets and mystics encourage us to be centered in the present—to see everything as if for the first time, to experience without the burden of past memories and future projections. When we do this, we operate from the quantum self. By contrast, ego responses are jaded responses in which we feel separate. In ego responses, we lose that extra-mundane quality of oneness, of no-separateness.

There are other differences between the ego and the quantum self as well. Quantum-self experiences have no ancestry, no identifying trace of previous memory. Thus these experiences have the same "alive" spontaneity for everyone. This is why the quantum self can be called a cosmic self. Identifying with the conditioned pattern of stimulus responses (habits of character) and the memories of past responses gives the experiences of the ego self an apparent local individuality. In other words, ego experiences are different for each of us.

But conditioning does not completely define the ego. In the ego, we also have the capacity to be conscious of our own past experiences. By using this capacity, we reconstruct our memories to suit ourselves in different situations. In other words, we create masks or personalities for ourselves. Somewhere in the process, we become the most important of all the different programs that we operate for our functioning. We become simply hierarchical. And we experience this as a function of our own uniqueness and importance.

When we operate from the ego, our individual patterns of conditioning, our experiences—which are predictable—acquire an apparent causal continuity. By contrast, the experience of the quantum self is quite discontinuous. Moreover, our physical individuality is both structural and functional. But our vital and mental individuality are subtle. They are purely the result of conditioning and are therefore purely functional. We are all potentially capable of accessing all the possibilities of the vital and the mental worlds, but, as adults, we generally don't do it. We don't have enough time, for one thing. Instead, we identify with a conditioned set of learned patterns with which we explore the vital and mental worlds. These individual functional patterns we call our vital and mental bodies, respectively. The conscious identity we experience with our physical, vital, and mental bodies, along with their correlated content memories, is what comprises our ego.

Limits and Risk

Once we identify with the ego, we are determined and predictable. How then can we be free? How can we transform? How can we escape from ego bondage?

True, the sum and substance of conditioning is that, as consciousness progressively identifies with the ego, there is a corresponding loss of freedom. In the limit of infinite conditioning, this loss of freedom would be absolute. At that stage, the only choice left to us, metaphorically speaking, would be the choice between conditioned alternatives. This is not real freedom. In a world of infinite conditioning, behaviorism holds. This is the so-called "correspondence limit," in which any new science predicts the same results as the old science. This is one paradoxical characteristic of any new paradigm. If the paradigm is correct, it must have a correspondence limit. In some limiting conditions, therefore, it must behave approximately as the old one did!

But we never go that far down the path of conditioning; we don't live that long. Even in our ego selves, we retain some freedom. And a most important aspect of the freedom that we retain is the freedom to say "no" to conditioning, which allows us to be creative every once in a while. This is the essence of risk-taking.

Risk-taking is liberating because, in a sense, it is the quantum self operating within you. So we must never be afraid of risk. And there is experimental data to support what I am saying. In the sixties, neurophysiologists discovered the so-called P300 event-related potential that suggested our conditioned nature. Suppose, as a demonstration of your free will, you declare your freedom to raise your right arm and you proceed to do it. An EEG machine attached to your brain via electrodes will generate a P300 wave that allows a neurophysiologist to predict that you are going to raise your arm. So actions of "free will" that can be predicted are not examples of real freedom.

So are behaviorists right that there is no free will for the ego self? Are mystics right when they say that the only free will is God's will to which we must surrender? But this generates yet another paradox: How do we surrender to God's will if we are not free to surrender?

Neurophysiologist Benjamin Libet did an experiment that rescues a modicum of free will even for the ego self. Libet asked his subjects to stop raising their arms as soon as they became aware of their intent to raise them. He was able to identify a 200-millisecond gap between the two events—between thought and action. He could still predict the raising of the arm from the P300 wave, but, more often than not, Libet's subjects were able to resist their so-called free will and not raise their arms, demonstrating that they retained the free will to say "no" to the conditioned action of raising their arms.

When I shared this information about Libet's data with a friend, he said he was glad to know, even in his ego, that he had the free will to say "no" to conditioning, because he used to be a smoker. When he tried to stop, when all the health warnings appeared and smoking in public places was prohibited, he found he could stop his tendency to light up—but never for very long. It took him a long time to curb his smoking to a socially acceptable level while maintaining internal ease. "And I still have the tendency to light up once in a while," he told me. "If we have the free will to say 'no' to conditioning, why is addiction so hard to give up?"

My friend raised a good point—one that will take us to the subject of intention-setting and creativity—indeed, to the whole science of manifestation in the next chapter. Let me end this chapter with an example of quantum-self action and one of ego-self action. Again, it is Zen that tells the tale:

> Two monks were about to cross a muddy river. Although a high current was making the river muddy, it was not really very deep, and was quite fordable. Just then, a maiden appeared in a beautiful kimono that went all the way to her ankles. Naturally, the maiden was hesitant to step into the river lest her clothes be ruined. One of the monks asked permission to pick her up. When she nodded, he carried her across the river and put her down. The maiden thanked him and went on her way. The other monk soon caught up with the first and they both continued on their way.

After about an hour, the second monk spoke: "Brother, you did something very wrong back there, you know. We monks are not supposed to touch women, let alone carry them for as long as it took you to cross the river—a full five minutes, and you held her so close."

The first monk said: "Brother, I carried the maiden for five minutes, but you are still carrying her."

The first monk performed an act of kindness, responding to his intuition that the maiden was in need of help. When we respond to our intuition, we act from the quantum self. The second monk was thinking from his conditioned ego self and his judgmental mind. So he suffered.

CHAPTER 8

Free Will and Creativity

In my travels, I entertain many questions about how to live the quantum worldview. One question I receive frequently is how to maintain present-centeredness. Downward causation is a causal ability of consciousness that allows us to choose the actual experience that we have from quantum possibilities. But do we choose this in the moment, or is what is happening at any given moment something that we chose or created in the past—or even something we choose from what we anticipate will happen in the future? The truth is that we have a tendency toward conditioning, so this is one of the problems of being human and fulfilling our potential. We have a perceptual, operational aid called the brain that stores memory and, when this memory interferes with our perceptions, past responses influence our present responses. We also have the tendency to project the future from these same memories and that, too, influences our present experience. As the great romantic poet Shelly said:

We live before and after
And pine for what is not.

All this lack of present-centeredness would not be so bad if it did not interfere with our creativity. To be creative is to choose in the moment, but it is a challenge in the sense that we have to transcend our conditioned ego to fall into that immediacy of being. This requires a process. Without the creative process, consciousness has the tendency to succumb to the brain and only experience objects and events through their reflections in memory.

Creativity, in other words, is not easy until you comprehend its subtlety. For creativity involves a process that includes preparation and some unconscious processing. Only then can a leap from the ego to a discontinuous creative insight occur. Ordinarily, thoughts are just parts of replayed memories and projections; they are therefore continuous. Only after a discontinuous new insight comes can you manifest a product that everyone can see as new—a new poem, a new technology, a new song, or a new *you*.

If you want to change your life today—to make it radically different tomorrow—you must engage in the creative process. This process requires the ability to respond without sifting through past memories. It also requires cohesiveness of intention, and purposefulness. You really have to wake up to the fact that you are not a machine randomly responding to chance events in the world. You are actually a purposeful, embodied consciousness.

The universe has a purpose; it evolves in order to make better and better representations of love, beauty, justice, truth, goodness—all those things that Plato called archetypes. When you wake up to this purposefulness, you become focused. If you don't tune in to the purposefulness of the universe, it all seems meaningless and you risk becoming hedonistic—you explore things that are pleasurable and avoid things that are painful. Your life will be driven by ordinary dreams—a big house, an expensive car, and other physical and material pleasures. But the real American Dream is about the pursuit of happiness, not pleasure. What's the difference? Too much pleasure always ends up in pain. But have you ever had too much happiness?

Freedom and Intention

We forget that it is life, liberty, and happiness that we seek. And liberty ultimately includes creative freedom. Without creative freedom, it means little. If liberty is simply limited to the freedom to choose the flavor of ice cream I want, I can do without it. I don't mind eating chocolate ice cream every day. But we seem to have lost touch with the necessity of *creative freedom*. Today, we are faced with crises that will require innovation and creativity to resolve. So people are talking about creativity again. But we need more than talk. We need an entire paradigm shift, a fundamental change in worldview. We have got to shed our very myopic materialist worldview and begin to live in a quantum world, the real world.

People often tell me that they want to change. But making changes is not a simple thing. We are not material machines. We can't just push a button or adjust a setting to invoke change. We are human beings and our creativity—our ability to create change—remains latent when we succumb to our conditioning, when we limit our lives to mechanical responses to what has occurred in the past.

To escape conditioning, we have to pay attention to our intuitions; we have to learn the art of *intention*. Also needed is unconscious processing, which requires focused purposeful preparation and patience preceding it. We have to allow time for things to gel in the unconscious in order to achieve new insights. Even when we get a discontinuous insight—a thought that has never occurred before—we still have to manifest that insight into the world. That new manifestation changes our perspective and represents a tremendous accomplishment of transformation in the way we sort things out in the world. This is not easy. On the other hand, it is not difficult either.

We have experimental data that shows the power of intention—data that most scientists ignore. But science is very segmented today, with each field or discipline operating within the confines of its own assumptions. Psychology has become an almost completely behavioristic and cognitive science as far as academia is concerned. Biology is chemistry, the biologists say, dismissing things like human intention. Physics—with the exception of quantum physics, with its consciousness-based interpretation—passes over the power of consciousness and intention in favor of mechanical laws and forces.

Ironically, it is nonscientists like Lynn McTaggert (*The Intention Experiments*, 2007) who are doing something to prove the causal efficacy of our intentions. Old-paradigm scientists continue to ignore the anomalous data of parapsychology, while the debunkers among them whisper that McTaggert is not really a reliable scientist. In fact, there is an entire industry of debunking magazines and journals that materialists publish regularly to discredit parapsychology. Other than these efforts to degrade it, mainstream science hardly pays any attention at all to this developing science based on the primacy of consciousness.

Parapsychology is based on the principle that consciousness chooses out of quantum possibilities to actualize the events that we experience. This principle is potent with possibilities for solving problems that are unsolvable under the materialist approach—problems relating to our health, our creativity, and our well-being. It is extremely important that we bring this new interpretation of quantum physics directly to the attention of the public. That is why quantum activism is crucial.

Dis-ease and Well-Being

What role does consciousness play in our health and in the manifestation of disease or sickness? Are you sick because your body is sick, or are you sick because you are ignoring your consciousness and the role it can play in healing?

Allopathy—modern medicine based on materialist science—is based on the idea that we only have physical bodies. All illness must, therefore, be grounded in the malfunction of an organ—for example, an organ impacted by a germ. But this is not only false in concept; it also conflicts with what we actually see. There are diseases like chronic fatigue syndrome that occur even when all the physical organs are functioning properly. Yet the patient constantly complains of pain. Where does the pain come from? Empirically and experimentally, the attitude that the physical body is the only place where disease can arise is, at best, incomplete. In the quantum worldview, we know better. Disease (dis-ease) can also come from our vital, mental, and intuitive bodies. This is also true in older spiritual traditions that were not so tied up in the materialist prejudices that govern science today.

In quantum science, we posit that consciousness has four compartments of possibility, each responsible for a different kind of experience. There is sensate experience, in which we sense the physical. And there are feelings, through which we experience the vital energies of the vital world. And we think; we have a mental cognitive facility that derives from the collapse of quantum possibilities of meaning in a mental world. And finally, we have experience of the archetypal world—a world of spiritual values that, when collapsed, gives us intuition. This is a world that is so subtle that we still debate whether it really exists.

So we live in a multi-verse of experience—four kinds of experience, derived from different worlds of possibility of consciousness. In the quantum view, we have a body in each of these worlds—a physical body, a vital body, a mental body, and an intuitive body. In fact, even consciousness (the whole) can be considered a body (called the causal body). Older spiritual traditions call it the "bliss body." Any of these bodies can become diseased. Any of them can cease to work properly, or be improperly accessed or used by us. There are alternative systems of medicine that claim disease occurs when these bodies are not performing correctly. Or perhaps they are not acting in synchrony. This, they say, is especially true for chronic diseases. Unless you remove the defect from these more subtle bodies, unless you bring back synchrony and harmony, you cannot remove the physical symptoms permanently; they will come back.

This is where allopathic medicine is completely helpless and hopeless, because it only takes care of physical symptoms. The symptoms return and must be treated again—generally through pharmaceuticals and invasive procedures. But allopathic drugs are actually very harmful to our bodies. They are poisons, and they can eventually cause serious problems in other parts of the physical body. So, in the name of healing, allopathic medicine uses potentially fatal substances.

The real answer to chronic disease is to treat the vital body through vital-body medicine systems like Ayurveda and Traditional Chinese Medicine, and mind-body medicine that brings the mental and the vital bodies back into balance and harmony within themselves, and into synchrony with the physical body. Pain and disease treated in this way will be healed for much longer, if not permanently.

Some people call these subtle-body medicine systems "vibrational medicine" or "frequency medicine." But these terms are imprecise. I prefer to call them vital-body medicine, or mind-body medicine. In my book book *The Quantum Doctor* (2004), I present an integration of alternative medicine practices and conventional allopathic medicine that removes the ambiguity of terms like "vibrational" or "frequency" in this context.

Vital-body medicine includes traditional systems like acupuncture, Ayurveda, homeopathy, and chakra-balancing. All of these medical systems treat the vital body. Mind-body medicine applies when the root cause of the dis-ease lies in the mind.

Cartesian Mind vs. Consciousness

Ever since Descartes, the mind has been interpreted in a very general way in the West—too general, in fact. It includes consciousness. It also includes what we normally call mentation—the mind as a function of thinking. Because both thinking and the thinker are included in this interpretation of the mind, there has been enormous confusion. When I talk about consciousness, I mean consciousness that includes the experiencer of the thought, as well as its object; mind now denotes the place where the objects of thinking reside.

The mind is the place where thinking takes place. More explicitly, it is the compartment of consciousness with whose help consciousness settles on the meaning aspect of the world. The mind helps consciousness give meaning to the physical world and to other objects of consciousness, including itself. The mind is thus the giver of meaning.

Many people think the mind is contained within the brain, that it is a product of the brain. But how can this be true when the mind processes meaning, whereas we have already shown that the brain is a computing machine that cannot originate meaning? Many people also think the mind is contained within the body's electromagnetic field—within its aura. But the mind is completely different from anything material, even from the biophysical electric body—the aura—which is itself a recent and exciting discovery.

The mind is the domain of meaning. Mathematician Roger Penrose has proven mathematically that computers cannot process meaning using step-by-step algorithms. In other words, it is fair to assume that meaning is outside the purview of the material world. But if it does not belong to the material world, there has to be a world where objects are objects of meaning. And this is the world that I call the mind—a world that traditionally was so called, until Descartes collapsed mind and consciousness into one. Everything internal became mind in Descartes' terminology, and that has caused enormous confusion in Western philosophy. Moreover, in scientific materialism, the mind is translated as internal to the brain, creating even more confusion.

In a recent interview, a journalist who was sympathetic to my point of view said: "Most people are under the impression that the brain is the thinking mechanism. Personally, I don't believe that to be true. I believe that the brain is the processor of the thinking mechanism. What are your thoughts on that?" Well, I didn't have the heart to offend her. So I said very diplomatically that we were using very subtle language and offered to rephrase the thought to make it more scientifically accurate.

I explained that the mind is the meaning-giver. The brain makes a representation of mental meaning. Once the brain has represented a lot of mental meaning, creating a software program of sorts, then certainly it is possible to say that the brain can process meaning, because it can process meaning that has already been programmed into it—meaning for which it has the software. And we do use this mental software—namely, our memory—most often in our thinking. We don't usually give new meaning to virtually the same everyday experiences; we don't usually process new meaning and engage in creativity that requires that we go outside of what we have memorized or what we remember with the brain. This is where the confusion lies.

And materialists, of course, take advantage of this confusion. They identify the brain as mind, thereby negating creativity completely. The proof that they are wrong lies in the fact that creativity is a very well-established experience that we all have. With creativity, we certainly can change the world—we can "disturb the universe," as physicist Freeman Dyson once said. Therefore, creativity has causal efficacy, no question about it. This proves that the brain cannot possibly be mind, because mind is needed for processing creative meaning.

"If the brain is not the source of creativity and intentionality, what is?" the journalist asked.

I explained that the source is consciousness itself. The experiencing subject that we become in a creative experience—that creative quantum self we sometimes call the holy spirit, the spirit in us, the spiritual in us—is the self that knows that nonlocal consciousness is the source from which creativity and insight come. And this insight comes in the form of new meaning.

For a long time, I went on, science has neglected its main objective of explaining what is the purpose of being human. In quantum science, we have discovered that purpose, which is to pursue, to explore, and to discover the soul—the archetypal or supramental body. Science has ignored the soul, ignored meaning. Because we talk about the mind as synonymous with the brain in our materialist culture, we have become extremely narrow in our attitude toward meaning in our lives. Day by day, our society has become more and more mundane, more devoid of meaning. We have become so brainwashed by the half truths of materialist science that we have completely forgotten about new human potentialities and we just go on repeating the same experiences.

So it is imperative that we recognize the paradigm shift that is taking place within science and bring it to the attention of ordinary people. However, at the same time, we must remember that all of us, ultimately, are part of the whole that I call quantum consciousness—what other traditions have called God. We have, in potentiality, the same potency as God. Although temporarily, we may be taken over by one cultural aberration or another—by self-imposed limitations, by conditioning. These are definitely not permanent states for us. We have gotten stuck in mistaken worldviews many times in our history—World War II and Hitler, for instance. But wars, violence, and a corrupted climate do not reflect all there is to human consciousness. It goes far beyond that. Materialism is like an epidemic disease that has to be healed. And quantum science can be part of the healing.

The journalist smiled and said: "That was a long, passionate speech. I know that Dean Radin's experiments show that consciousness can affect where the balls in a pinball machine are and where those balls will drop in that machine. So it has to be true that, if people would just understand that they are the source of creativity and intention—they are the oneness, the God source, the God consciousness—if they would just project thoughts of agape or positive intent for mankind, they could make something positive happen."

To this I added that we also have to discover where we fall short. And we have to recognize why our intentions fall short, why they become so narrow in terms of their potentiality and keep us from transforming to that bigger consciousness. The fact is that evolution has given us negative emotional instinctual brain circuits that limit our consciousness to a negative emotionality. Even when we have positive intentions, we are also thinking: *What is in it for me?* So we never get beyond the positive thought to a positive intention in our hearts. And we never act on these feelings to create positive emotional brain circuits. We never feel expansiveness in the region of the heart that Easterners call the heart chakra.

Seeing her nod in agreement, I lamented that we have forgotten what mystics call the journey toward the heart, especially in the technologically and economically advanced West. We suppress feelings, thus losing touch with a very easy way to expand our consciousness—namely, bringing the energy in the head down into the heart. When we learn to do that, unconditional love comes to us in a very natural way. When we feel the heart expand, our intentions have greater potency and a much better chance of actualizing in the world. When we intend world peace with an expanded heart, it has much more of an effect than if we just intend it by thinking about it, because when we are thinking, we are already narrow and self-centered. If we try to bring world peace by changing others and not ourselves, we will fail. We have to do both. We have to change ourselves as well as others.

She nodded in agreement. But her next words showed confusion once again: "I think that anytime you think a thought, it causes neurotransmitters to fire and something happens. And I believe that, when you go to the position of your heart—with love in your heart, and not just in your brain, in your mind—it causes neurotransmitters, neuropeptides, hormones, and other chemicals to be released into your body. And this causes your heart to be filled with blood, and that feels good. But I also think that it changes your resonant field—what you call the correlated morphogenetic field. So, if your resonant field changes to a more positive resonance, it can't help but affect the resonant frequency of the entire world. The effect may be subtle, but it is there just the same. Does that make any sense to you?"

Again, I tried to gently push her toward a quantum worldview. I explained that, in quantum physics, this happens because of a phenomenon called quantum nonlocality through which our intentions can become correlated because we share the same consciousness. Consciousness is the ground of all being. We thus all come from that one consciousness. Moreover, there is now laboratory proof that this concept of one consciousness is valid. Although in our ego state we may not experience it, there is a nonlocal consciousness, a connectedness without any signals, a signal-less communication that we can share. With it, we can influence people, even without the intermediary of electromagnetic or sound waves.

The Dilemma of Choice

Then, my journalist friend asked the crucial question: How does all this correlate with free will? I told her that free will comes from that ultimate causation that is beyond all material causation—downward causation. Quantum physics is the physics of possibilities, and consciousness is needed to choose from these possibilities. That choice, when made freely without past conditioning, is what we call free will. We do have free will, but it occurs in a higher state of consciousness—in that consciousness that some call God and I call quantum consciousness.

Many people are not particularly conscious, because they don't really use the freedom of choice that we can have through an evolved consciousness. In other words, we lead a zombie-like existence as more-or-less conditioned beings. But it is within our power to escape this. And we can start by saying "no" to conditioning.

I could see that she was becoming enthusiastic. "So I can say that I'm going to the grocery store tomorrow at three o'clock," she reasoned. "That's my free will to choose. But, in actuality, I have absolutely no way of guaranteeing that, even though I fully intend to be there at three o'clock. I might be stuck in traffic. I might have an accident. I might get a phone call that detains me for an hour. So at the source of consciousness, do you believe that free will exists for us to do whatever we choose?"

I laughed and replied that there is no guarantee, because your intention depends on other factors that are, in turn, dependent on how other people are exerting their own free will. Then I explained that the words "free will" can be used in two ways. Free will can mean choosing between conditioned alternatives. In other words, your conditioning contains the possibility of transport and the availability of a grocery store, and you have access to both at any time you choose. So, in effect, you are choosing between alternatives that you have already experienced and remember. But that is not unrestrained, totally free will. Materialist scientists have very good models that explain conditioning as neural nets that are "grooved" into various pathways in the brain as they respond to repeating stimuli. Because there is more than one grooved pathway, you can choose between different responses. But this choice is not totally free, because it depends on conditioned learning that your brain has already accomplished. You don't require the free will of God consciousness to explain that.

One of my favorite examples of this is when you agonize over an ethical decision. Sometimes ethical decisions are so tricky, they require so much finesse, that the choice becomes agonizing because it is so ambiguous. I often have this debate with well-meaning people, because sometimes I like to make radical statements just to encourage them to think more deeply. For example, I said at a lecture that all conscious people, transformed people, reject truth-telling in a literal sense. People got very upset and accused me of suggesting that people who become conscious are liars. But consider this ethical dilemma.

You see someone running. In the next moment, you see someone running after that person with a gun. If the person with the gun stops and asks you if you saw someone running, you may not immediately respond, "Yes, he went that way." Instead, you hesitate and say something vague or noncommittal, because you don't want anyone to be shot. But if you realize that the person with the gun is a policeman, you may change your mind. That moment is a moment of creativity—the moment of your free choice based on a gut feeling. You didn't succumb to your conditioning by immediately answering when the person with the gun asked you something in an authoritative voice. This was an act of freedom, of resisting your conditioning. You hesitated, and that was the creative opening, the opening for a creative decision. In this kind of situation, we can see that we have causal efficacy, and we have to bring it into manifestation. We do that more often than not when we are conscious, when we are a little transformed. Explained this way, it makes sense that sometimes we may have to tell a lie to save a person.

In its essence, I told her, free will is about creativity. When we are creative, we exercise freedom, because we choose something that we didn't know before—something that is totally new. So real freedom is exerting a choice that cannot be predicted—one that was not experienced before, that is totally new—something over which the ego has no control. The freedom to choose freely among our own conditioned alternatives is important, and we fight for it. We fight our parents for our choice of ice cream flavors when we are children. We fight them to make our own choice of college when we are young adults. When Patrick Henry said, "Give me liberty or give me death," he was expressing that kind of freedom. It is important, but it is not the ultimate freedom; it is not creative freedom. It is not the freedom to create something that is completely new, although it can be an important step toward that.

She thought about this for a moment and then said: "So it is more as if we are two-year-olds sitting in the back seat of a car strapped into a car seat with a little yellow steering wheel. We think we are driving the car, but we really aren't."

"Exactly," I said. "We *think* that we are driving. But it is only when we take on creativity that we really begin to drive. Remember though: The human body-mind is not ready to be creative all the time. Even people like Jesus and Buddha were not in their Holy Spirit world all the time. And when they were in the physical world, they spoke in parables—in words and in languages that we don't always understand.

Manifesting Choice

Then my journalist friend came back with a wonderful question: "What are your thoughts on the manifestation of intent into physical reality? The average person doesn't believe that intent is powerful. They want to know why they can't manifest whatever they want—like a Rolls-Royce, for instance."

Of course, this is one of my favorite topics. Movies like *The Secret* make a fortune selling simplistic notions about manifesting choice in the physical world, although the quantum worldview does indeed suggest a science of manifestation if taken to its logical conclusion. "In fact, in the 1970s," I told her, "there was an organization that taught people how to manifest

Rolls-Royces. If they failed in that, they promptly began teaching them how to manifest parking spaces in busy downtown areas."

My friend laughed, and I continued more seriously. "It all comes down to the old problem of the ego—the narrowness of consciousness that keeps us from manifesting our intention. We identify the ego's limitation as a limitation of intention. Intention has the potency of God consciousness, but we have to intend in the right frame of consciousness. If we work within the narrowness of the ego, our intention is not going to have any effect on the cosmic consciousness where such manifestations are open as a possibility. If we intend from heart consciousness, however, we become somewhat more expanded and our chances of success increase."

I continued by explaining that if we can occasionally access mystical states of consciousness, as Jesus and Buddha did, and become really synchronous with cosmic consciousness, then our chances of manifestation become even better. It depends on how cohesive in consciousness we want to be and how much expansion we can attain. In expanded states of consciousness, we only intend good for everyone. We don't work for individual gratification of the material kind. Our selfishness goes away. But this frightens some people who only want their selfish goodies and the satisfaction of the senses. So, as a collective, we have some growing up to do. We are still children in terms of maturity of consciousness.

We really must learn the subtleties and the enormity of the potentiality that exists within us. Collectively, we have not even begun to expand into mental creativity, into the discovery of new meanings. Great artists, great scientists, and great philosophers may do it. But the rest of us are quite satisfied with experiencing a very narrow range of mental meanings. So we toy with the idea of manifesting something that did not exist in the material domain before—like a Rolls-Royce. But that's a huge order. It goes completely against the grain of materialist physics, because it involves creating new matter out of nothing. What we need to explore is a much simpler kind of creativity—the creativity of mental meaning. Our ancestors knew this. So when they prayed to God for something material, they knew God's grace would manifest through the benevolence of another human being.

We have a long way to go, I told her. But that does not mean we are stymied. As the Chinese proverb says: A ten thousand-mile journey begins with the first step. We have to learn to be creative—first with mental creativity, then with our vital energies, and finally with creativity at the material level—which is tantamount to what we call a miracle.

CHAPTER 9

Involution and Evolution

Attendees at my workshops often ask me what I mean by the ground of all being. Do I mean God or whatever someone may call that deity? I tell them, technically yes. But I prefer to define God as the *creative agent* of the ground of being. The ground of being, philosophically speaking, should be thought of as eternal and inclusive of all possibilities. Anything that is eternal is outside of space and time, so nothing can happen in eternity that is inclusive of all possibilities. For creativity to come into the picture, for things to happen, there must be some limitation. So God is not exactly the entire ground of being, but is rather the creative agent of the ground of being, after some limitations have been imposed.

Before God

So what is the ground of being if it is more than God? The ground of being is sometimes called the God Head in spiritual traditions. It is eternity, inclusive of all possibilities, that is always there in the background. The buck stops there; otherwise we go on asking: What was before? What was before?

It's like the old conumdrum about the Big Bang theory. What came before the Big Bang? Materialists posit an explanation they call "inflation theory," which projects a time before the Big Bang. But this begs the question of what came before the inflationary epoch. Spiritual traditions don't do very well on this question either. The story goes that St. Augustine was once preaching about how God had created Heaven and Earth when one of the faithful asked him what God was doing before He created Heaven and Earth. After a pause, Augustine responded: "God was creating hell for people who ask such questions."

But what *was* God doing before he created Heaven and Earth? What *did* come before the Big Bang? The only sensible answer is that, before the Big Bang, before God, before anything we can think of, eternity is always present. And this is the ground of being. The "buck stops" with eternity, because nothing can come before it. Eternity has no past, no present, and no future. It contains all possibilities—everything that is, or was, or ever will be. Eternity is timeless. To bring time into the picture, to bring creativity into the picture, there must be limitation. We must bring in laws involving quantities that vary with time—not manifest time, but with time as parameter. We must bring in meaning—the idea that there are definite templates of biological being with which consciousness can work. Only after all of these limitations are introduced can matter be created and manifestation take place. Matter then makes representations of what went before it—representations of the templates or blueprints of biological functions, representations of mental meaning, etc. (see Figure 11).

But representation and meaning require an observer. The original observer is theorized to be a one-celled creature from which we all evolved. More than four billion years ago, this original life manifested. That was a monumental event. Before then, everything was just possibilities from which consciousness could choose. There was no manifest consciousness, no

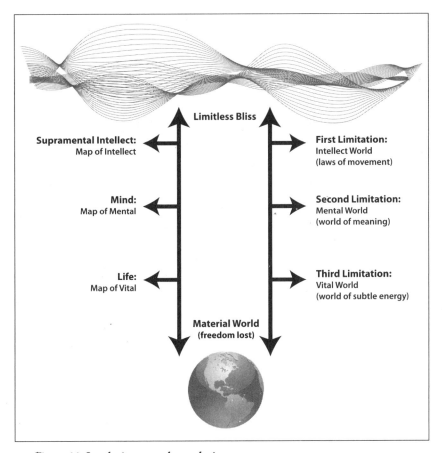

Figure 11. Involution precedes evolution.

manifest observer, and no choice. It was a possible universe, so to speak, with no actuality. In that moment of first observation—in that moment of what physicist John Wheeler called "the completion of the meaning circuit," or more accurately, the completion of the tangled-hierarchical life circuit of the first living cell—the whole universe was created, going backward in time through delayed choice. Of course, we cannot rule out the possibility that there are other planets besides ours where other such one-celled creatures have arisen. But with that caveat, I can say that, before life appeared on the scene here, everything was just quantum possibility.

It is difficult to comprehend at first what we mean when we say that the universe was created going backward in time from the point at which the first living cell sprang up about four billion years ago. The Big Bang, our latest calculations show, probably occurred about thirteen and a half billion years ago. But this creates a conceptual problem. The heat generated by the Big Bang would have precluded life, so no manifest observer could have existed, and therefore no choice was possible. So it must be that, when the first oberserver made the first choice, it was a delayed choice. When the first living-cell consciousness chose, four billion years ago, all the things that went before that event—the atmosphere on earth, the sun and the solar system, supernovas, first-generation stars, the galaxies, the Big Bang itself, the entire causal lineage—manifested retroactively right at that moment. Even space and time, which were parameters in potentiality, were created with it (see Figure 12). And time was created with a linear chronology that, logic tells us, it must have in order to come into manifest being. So our choice *now* precipitates the actuality from possibilities *then*, going backward in time.

Evolution and Purpose

Let's talk a little bit more about evolution and consciousness. Did our world evolve or did it appear spontaneously as the creationists say. Or is it a combination of both?

For a long time, spiritual traditions and religious dogma taught that there is no evolution, because God created everything all at once. This mistake arose from the assumption that the universe is not lawful. Our experiences are so full of chance

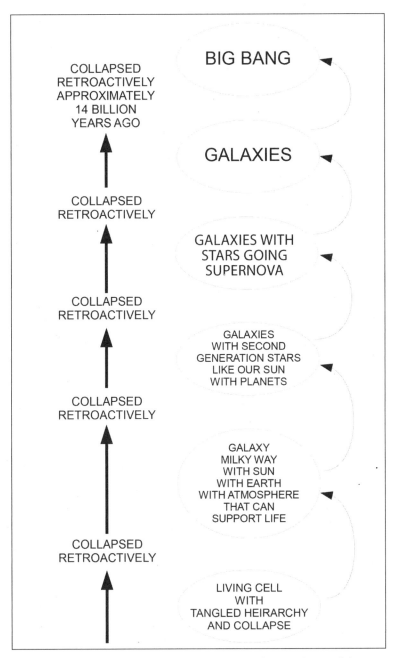

Figure 12. The retroactive collapse of the Big Bang; the completion of the tangled hierarchy of the first life; delayed choice and collapse. (Illustration by Terry Way)

happenings that it took humanity a long time to fathom that the universe *is*, in fact, lawful and that God works within those laws. Once science established the lawful operation of the universe, it questioned whether there were also laws that circumscribed manifestation. Science no longer needed to assume that manifestation occurred arbitrarily, all at once, so it challenged the idea put forth in Genesis that God created the world.

Scientific evidence in the form of fossil records now clearly shows that evolution occurred. But here is what biologists try to ignore: evolution is progressive; it has an "arrow of time." Single-cell creatures gave way to multiple-celled animals, which gave way to invertebrates and vertebrates, which eventually gave way to primates and human beings. But there are many anomalies and many missing intermediates between the stages of macro-evolution in the fossil record that are not accounted for in Darwinism.

In my book *Creative Evolution* (2008), I address these anomalies. There is a metaphorical interpretation of Genesis—that creation spanned six days—with which we can avoid the idea of instantaneous creation and make room for evolution. In the same way, Darwinism can be modified under the light of the primacy of consciousness into a new theory that explains the arrow of time and the missing pieces in the fossil record by making room for creative quantum leaps.

According to Darwin's theory, the fossil record should show a continuous development from one species to the next, accounting for all significant differences. But this is not the case. The fossil record is not continuous. There are gaps in it, especially when we consider macroscopic groups higher than the species. For example, between amphibian creatures and reptiles, Darwin's theory says there should be thousands of intermediates.

In actuality, however, only a few, about sixty, have been found. What do these fossil gaps mean?

Here is where quantum science can help to explain. In the quantum view, you don't have to assume that evolution proceeded only through continuous mechanisms—chance mutations of genes and natural selection from the mutated genes selecting those that are beneficial for survival. Instead, we can posit evolution as the evolution of consciousness. When we look at evolution from the point of view of consciousness, we can invoke creativity. We can argue for a conditioned continuous action, as in traditional Darwinism, but also for creative action—for quantum leaps—in the evolutionary process.

When we become conditioned, things become continuous. Have you noticed that our stream of consciousness is quite continuous? But creativity occurs intermittently and that is what enables us to start new phases of life. Biological evolution proceeds in the same way. There are some instantaneous changes that take place—quantum leaps—and these explain the fossil gaps.

When we introduce consciousness into the equation of evolution, what drives intention? How does information about intention come from the ground of all being into matter? How does it get to the material level? Through signals? Remember that intention is not a cause. A better question would be: What is the driving *purpose* of the universe? How does conscious *purpose* get into matter? And this is where biologists have to open their minds to quantum nonlocality. For the fact is that, in quantum consciousness, no signals are required. Quantum consciousness affects matter, material possibilities, through nonlocal downward causation—through choice.

But why ask all these questions at all? Many people think that God is perfect, therefore the world should be perfect. But here is the catch. The world is not perfect and people wonder why. Evolution gives us the way out. The idea of evolution is that we begin as imperfect beings, but we try to become perfect—we try to transform ourselves; we try to evolve as spiritual beings.

Then why did God manifest the universe? Why not stay in perfection forever? Because in the unmanifest, in eternity, nothing happens. There is no experience. Things happen because we want to attain perfection eventually, but we want to attain it in manifestation, in experience. And this is the purpose of manifestation. Scientists accept that there are causal laws of the universe; they also need to accept that there is purpose.

The Anthropic Principle

There are eight major archetypes that we recognize—archetypes of value that drive our social evolution. These are love, beauty, justice, truth, self, wholeness, abundance, and goodness. We have to be creative to manifest these archetypes. Can we love by just wishing it? Can we be just by just wishing it? No, we have to struggle for it. We have to evolve toward it. We have to be creative. The whole idea of evolution is that, initially, we begin as not very loving, or not very just. Initially, we are not very perfect. And how could we be? Consciousness had to create life within the limitations of scientific laws that take advantage of contingencies. But eventually, we can evolve toward becoming loving beings; we can evolve toward justice, or truth, or goodness. We can evolve toward perfection.

Of course, sometimes we backslide. In society today, we don't see very much justice. We make progress like the proverbial monkey going up the bamboo pole—one foot up and three feet back. But fortunately, the "taste" of a just society remains with us, and eventually we catch up again. And when we catch up, we go a little bit higher. That's the way of evolution. So we need not feel too disheartened.

That the world is designed to move toward establishing manifest embodied consciousness is called the anthropic principle. Anthropic means "human" or "human existence." Principle can be defined as "law." Thus the anthropic principle purports to be the law of human existence.

The anthropic principle is a good example of delayed choice. Without delayed choice, it would be unexplainable. With the anthropic principle guiding it, the world evolves in such a way that, eventually, a manifest embodied consciousness will come about. In quantum physics, manifest consciousness is what makes a manifest world possible. Only when manifest consciousness comes about can the world manifest. And then the circle is complete. The world needs consciousness to manifest, and consciousness needs the conditions under which consciousness can manifest. This circularity is what drives the anthropic principle. We need the universe; the universe needs us. The universe cannot manifest without the observer. The observer cannot manifest without the universe and its evolution.

CHAPTER 10

A Tale of Two Domains

Now let's go the heart of how quantum physics and the theory of relativity—a theory of time and space—can be compatible. First, let's consider the idea of signal-less communication—what, in quantum physics, we call nonlocality. On the face of it, relativity theory seems to conclude that there cannot be instantaneous communication, there cannot be instantaneous quantum leaps. In relativity theory, there is always a finite speed of communication limited by the maximum signal speed, which is the speed of light. Now let's look at the quantum concept of two domains. Quantum theory defines the domain of potentiality as a nonlocal domain that exists outside of space and time, beyond the jurisdiction of the theory of relativity. Thus, what appears to be discontinuous in the space-time domain happens only because there is signal-less communication that occurs via the domain of potentiality. In other words, potentiality and actuality are mutually interdependent phenomena.

Take, for example, the idea of quantum correlation or entanglement that occurs between objects when they interact. Suppose we measure two entangled objects that are millions of light years away and find that, when object A is polarized in one direction, object B is always polarized in the same direction. This is a case of correlation, or entanglement.

The nonlocal connection between these two objects, which is a potentiality, needs a trigger to be activated. In other words, every object is potentially connected through nonlocality with every other object in the universe. But, in order to actualize that nonlocal connection, we have to entangle the two objects, or correlate them, through some local interaction.

This may seem confusing, but think about the potentiality domain, the domain of oneness. We know that is similar to the concept of the unconscious in psychology. And we know that the personal unconscious is rooted in the physical body, in brain memories. But Jung discovered that the collective unconscious is also rooted in memory—in the ancestral collective memory that we collectively suppress. But the unconscious itself, which is identical to the domain of potentiality, is bigger than the sum of those two parts. It is much bigger than either the personal or the collective unconscious. And that "extra part," the unconditioned new quantum possibilities, we will call the quantum unconscious.

From Subject to Object

Werner Heisenberg, who was a co-discoverer of quantum physics, first made this connection. Think back to our thought experiment of chapter 2, in which we hypothetically released a free electron in a room and tried to measure where it was. (Here, "free" means free from all forces.) Heisenberg asked: What actually happens when potentiality changes into actuality? When we know only potentiality, all we can say, all we can calculate, all we can know, are probabilities for an object to be at various possible locations. Our knowledge about the object is vague. But when we convert the object into a particle of actuality through the event of collapse, we know exactly where it is. So Heisenberg said that the change in an object from potentiality to actuality is a change in our knowledge about the object. Now we have to ask the question: What is the vehicle that brings us that knowledge?

Consciousness, as we have seen, is the vehicle with which we know. So what Heisenberg is saying is that the change of potentiality into actuality is a change in consciousness. And so this domain of interconnectedness is the most general form of consciousness we access. Before (and beyond) the ego, we *are* this vast, interconnected ground of being that is consciousness. It is an unconscious that we all share. It has no personality, no self of the kind that we experience in the manifest world.

We can express this in a diagram showing an individual self and its relationship with consciousness outside space and time (see Figure 13). Potentiality comes first, then change or collapse, then the subject-object split that results in actuality. This establishes the manifest observer, who acts as a trigger for the domain of actuality.

The individual observer is the current manifest consciousness in association with the brain appearing as the subject—call it the quantum self—of a spontaneous experience. Some of the potentiality becomes the object of that experience. The brain is involved in this collapse, because no change of potentiality to actuality of any experience can take place without the tangled hierarchy in the observer's brain.

Let me repeat that: *No change can happen without the observer's brain.* Consciousness is represented—embodied—as the subject identifying itself with the brain. Each time a child is born, its brain provides the tangled hierarchy manifesting the reality that the child experiences. And the child becomes an observer as a separate self, although a baby is not quite aware that it is a self separate from its environment until it is about

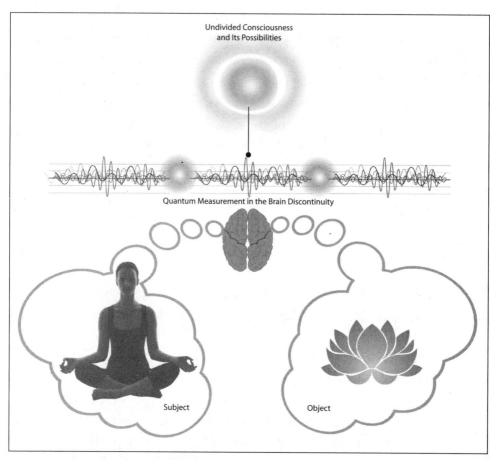

Figure 13. How the one becomes two; the subject-object split upon collapse.

a year old. At least a year-long memory has to accumulate to support this perception of separation. Only when the spontaneous experience of the quantum self is reflected in the mirror of the accumulated memory do we begin to discern the ego self as separate from the environment.

Hinduism identifies these two realms as *purusha*—potentiality of subject-consciousness—and *prokriti*—potentiality of objects. From this, upon what we would call quantum collapse, is born the manifest awareness of a subject looking at objects. In

Christian metaphor, purusha, the potentiality of consciousness, is God the father. Prokriti, the potentiality of objects, is the mother. God (purusha) impregnates (via downward causation) the mother (prokriti) to produce (immaculate conception) the manifest subject-object awareness of the quantum self (the Holy Spirit) and the objects of experience. Finally, conditioning makes the quantum self into the ego (the Son).

In the Hindu monistic philosophy of *Vedanta,* purusha and prokriti together make wholeness—*Brahman* in Sanskrit. This is the Godhead of esoteric Christianity. Likewise in Buddhism, potentiality is recognized as "no thingness" (nothing-ness). They call it *shunyata,* which transcends both the subject and the object. Thus the quantum description of consciousness was anticipated millennia ago.

From Potentiality to Perfection

Before measurement, before observation, there is a potentiality of an object and the potentiality of a brain looking at it, both contained within consciousness as possibilities of consciousness itself. The change, which physicists call collapse, leads to two things—the collapsed brain develops a representation of consciousness (the subject) and a collapsed object, which the subject experiences as the object of experience.

Let's see how this applies to how a bottle of water emerges from potentiality. The bottle and the observer are "dependently co-arising." To begin with, the bottle and the brain are all one with consciousness in potentiality. Here, in manifestation they are separate—the observer is separate from the bottle. In ordinary ego experience, the observer is also separate from the whole of consciousness. So potentiality is the domain of unity, while the manifest is the domain of apparent separateness.

The unmanifest is wholeness. It is perfect. When you know this in your heart, you will have taken a major step toward enlightenment. So why do we need anything besides that wholeness? Why do we need this separate reality? Why do we need the bottle of water? Why did we create it? This, of course, is one of the fundamental questions that religions and spiritual traditions puzzle over. Hinduism calls the world of separateness "God's play." This satisfies in part, because you cannot enjoy eating an apple in potentiality. Some religions see separateness as original sin. Their prescription is to avoid it and keep your attention focused on unity. But they miss something with this trivialization of the manifest world. The manifest world has order. Part of the order is even subject to scientific laws.

So why does the whole—the perfect—create the appearance of separateness? The answer is that there is no experience in potentiality. When we say potentiality is beyond time, it also means that past, present, future, everything coexists simultaneously in potentiality. But there is no manifest time and no experience. So the manifest world is created to create experience—and eventually, via evolution, the experience of perfection, first for a few enlightened individuals and eventually for everyone. A little like what the science fiction writer Arthur C. Clark envisioned in his prophetic novel *Childhood's End*.

It is true that consciousness represented is not the same as consciousness in its original. But if we create a world that evolves, then it is possible to imagine evolution as a process through which we grow, becoming better and better adapted to manifesting love, goodness, wholeness, abundance, or any of the archetypes we mentioned before. And this is the new spirituality of quantum physics. It was already anticipated by two great thinkers of the last century—Sri Aurobindo and Teilhard de Chardin—who both saw the purpose of manifestation as bringing heaven on earth.

When the Big Bang happened, it began an expanding universe, and the world has kept on expanding ever since, becoming bigger and bigger. This world of reality emerged from the unmanifest—from wholeness, from perfection—according to quantum theory. But just before the moment of the Big Bang, what was the intention—the primary intention that set off everything? The intention was, in one word, manifestation. But then what was the purpose? Put simply, the purpose was to bring the perfection of the unmanifest (heaven) to manifest (earth). But for that, we need evolution. And what drives evolution? It is the purpose of evolving perfection. Q. E. D.

From Unity to Separateness

Manifestation, experience, requires a mechanism for creating separateness. The Big Bang consequently had to be followed by a lot of development in potentiality before the manifestation in actuality could take place. About a billion years after the Big Bang, galaxies emerged. Then stars. Then supernovae. Then, out of the supernovae's remnants, second-generation stars like our sun. And these stars had all the elements needed to create life on a planet that revolved around a star like the sun and drew energy from it. So then the planets had to evolve, in potentiality, an atmosphere that could support life. All that development in potentiality took about nine and a half billion years—also in potentiality. And then the first living cell came about in potentiality. Finally, the mechanism for creating appearance, tangled hierarchy, was there in potentiality in the structure of the living cell, and the environment was just right.

In a living cell, there are two important molecules: protein and DNA. The origin of these molecules is very mysterious. Without DNA, we cannot make protein. But without protein, we cannot make DNA. Together, they constitute a tangled hierarchy in the living cell. So now collapse can happen, giving us a manifest living cell that can distinguish itself from its manifest environment. There is also a morphogenetic field associated with the form and function of the living cell—the cell's blueprint. The collapse of the associated morphogenetic field gives the cell a feeling of aliveness, however rudimentary.

Notice the similarity to the concept of "dependent co-arising" discussed before. As it manifests, a cell becomes an entity separate from its environment; they arise co-dependently. And this process gives rise to the living cell—to life—and its non-living environment. The protein-DNA combination, along with its cell wall and some cytoplasm, becomes life, and all the rest of the manifest possibilities become the non-living environment. So tangled-hierarchical quantum measurement thus led to manifest life, while the rest, with no tangled hierarchy, became the non-living.

Let me be clear. Where there is no tangled hierarchy, we get a non-living entity. The tangled hierarchy part becomes special—what Humberto Maturana calls an *autopoetic* system—with self-autonomy and self-identity. This is because consciousness is represented in these systems as a separate identity, as a distinction. In the rest of the manifest world, consciousness cannot be represented, so it manifests as objects.

At the time of the Big Bang, there were no living creatures, so there were no observers to observe the process. This is another wonderful paradox to consider, and its resolution is crucial. The answer is that, not only can this choice of collapse happen now, precipitating an isolated event in this moment; a choice now can also precipitate retroactively, going backward in time, an entire chain of events that are prerequisite to this event now. So, when consciousness chooses with a delay, a time delay, collapsing an entire chain of potentialities and producing an entire chain of events going backward in time all the way to the Big Bang, the delayed choice creates time retroactively.

We tend to think in a framework of absolute time, so delayed choice is difficult for us to understand. But when you allow the concept of time to be flexible, created by consciousness itself via collapse, delayed choice starts to make sense. In other words, we could say that our current consciousness created the Big Bang. Of course, our consciousness was not always so current. The Big Bang occurred thirteen and a half billion years ago. The first living cell appeared about four billion years ago. So consciousness has been doing this—playing this cosmic game of manifestation—for only four billion years. The first nine and a half billion years was a development in potentiality. Then a series of delayed choices and retroactive collapses occurred. At that moment, four billion years ago, life was created. Going backward in time, the sun was created. And before that, the supernovae; before that, the first-generation stars; before that, the galaxies; and before everything, the Big Bang occurred. But the whole chain of events is the result of one event of downward causation.

Without the Big Bang in potentiality, there would be no galaxies or first-generation stars in potentiality. There would be no supernovae or second-generation stars in potentiality. There would be no planet like our Earth and no living cell in potentiality. So this chain of potentialities is essential to produce the actual event of collapse that produced life. And before this actuality happened, there was no space; there was no time.

Wow! Huh?

From Theory to Fact

All this theorizing has brought us to a point where what I have been saying can be experimentally verified. One means of verifying the theory of delayed choice—of how an event now can precipitate a whole series of events that actually existed before—has come from research on near-death experiences.

What happens in a near-death experience? Cardiac failure occurs. The patient dies. The patient is pronounced brain-dead. The rest of the body is still alive, but the brain is dead. There is no EEG signal. With recent progress in cardiology, however, cardiac surgeons can revive a patient provided not too much time has elapsed. Then the patient wakes up and says: "I had this experience. I went through a tunnel to the other side and met my relatives. And after that I met Jesus (or Buddha, or whomever). And then I saw wonderful light. And I heard a voice calling me back. And I encountered a wall. And here I am. Back where I was."

From a scientific view, the difficulty in understanding these experiences is tremendous. A brain-dead person cannot have an experience that involves meaningful thought. Experience of meaning requires a tangled-hierarchical system called the brain. But the brain is dead. Yet we must take all these experiences seriously, because many other people in similar circumstances have reported similar experiences, and because these events change people's subsequent behaviors—their whole future lives. So these events have causal efficacy.

How could these events have occurred? Quantum science would answer that there must have been a chain of events—a retroactive chain of events—that began when the patient was revived. In this view, near-death survivors are remembering the chain of events because it is now part of their memory. The events occurred, going backward in time from the moment of collapse, which must be the moment when the brain was restored. It's the same thing that happened with the Big Bang and the origin of life. All we can look at today is the memory of the Big Bang contained in the microwave radiation background that pervades the universe—radiation that was emitted at the time of the Big Bang, but has since cooled off. So the analogy holds. What we see today occurred billions of years ago, existing now as a memory.

Creation takes place moment to moment. The universe out there does not exist in concrete form. Galaxies, stars, and planets aren't just sitting out there. They all exist in potentiality, unless some living being has an experience that manifests them retroactively. You could say the Big Bang is being collapsed moment to moment as astronomers look through their instruments and detect the microwave background radiation. In fact, quantum physics throws out the whole Newtonian concept that the universe is always just sitting there in concrete form.

So what about the moon? Does it exist? In a paper in *Physics Today*, a physicist asked the question: Is the moon there when nobody's looking? And the conclusion is unequivocal. Quantum physics insists that there is a no moment-to-moment moon, only the potentiality of the moon. Of course, the potentiality is so close to actuality—the moon is such a large mass—that we always find the moon where we expect to find it. Because of its size, the possibilities of movement of the moon's center of mass are very limited. So the Newtonian description is approximately valid. And what about Neil Armstrong? What happened when he landed on the moon? Well, Armstrong was manifesting himself, and his manifestation was co-dependently creating the moon as part of his experience.

And what about me? I am sitting at a table with a computer and creating a book, and you the reader will be reading that book about a year from now. You are obviously in potentiality. Am I in the world of manifestation? Do I really exist in this manifest reality, all the time that I am writing this book? Indeed, no. All we can say is that I am jumping in and out of that potentiality and creating a reality so close to what we expect it to be that we falsely assume I am a fixture, sitting here continuously writing the whole time. The continuity is a creation of our expectation. Between periods of writing, I sleep, and I become potentiality. And truly, that's the whole idea. Once you understand quantum physics, you can never take the fixity of reality very seriously. Nothing is permanent. Everything is always changing. Everything is ephemeral, just as the mystics have been saying all along.

From Fixity to Change

Of course, not all quantum physicists perceive reality in this way. But most believe that there is a fundamental dynamism in the world. When we become creative instead of living in ego-fixity, we begin to experience these discontinuities in such a way that we begin to see change, not fixity, as the basis of the world. Fixity is, however, necessary as a reference point for change. Change and fixity—movement and stillness—occur at just about the same time.

The important thing is to understand the distinction between the immanent and the transcendent, the manifest and the unmanifest. This computer I am working on is immanent. It seems to exist continuously in three-dimensional space quite separate from me. If I push it, it moves. The apparent actuality in the material dimension is matter moving in three-dimensional space and taking time to do it separate from me. But in the transcendent domain of reality, there is no separateness. That is why we cannot touch it or perceive it as separate from ourselves. This domain is outside of space and time. It transcends space and time. In this domain, all space exists simultaneously; all times exist simultaneously. All is one. There is no separateness between subject and object.

You also have to remember that this potentiality exists eternally. Moreover, there is an intention in potentiality for manifestation and that means that the anthropic principle is built into it at some stage of involution. And the world as we know it is created. And then, perhaps, after many billions of years, that world falls apart and cannot support life any more. And then, the intentionality of the potentiality creates another world.

This time around, consciousness chose the physical laws that governed the evolution of this particular universe. But according to quantum theory, there have been many universes with different physical laws governing each one. Some of them never collapsed because they ended too quickly. The potentiality of life never developed in actuality, so there was no manifestation. But at least one of these universes—our universe—must have been very well-adapted to make room for life. The physical laws must have been extremely fine-tuned in order to make this tangled hierarchy possible. If you change the physical laws even a little bit, there will be no tangled hierarchy and no manifestation.

I hope, like me, you have started to feel the dance of Siva in all this creation and destruction. When I was a materialist, I thought that things happened in linear time. But that is not so. Everything is processing, but nothing is "happening," in potentiality. Only in actuality do things happen in a linear way. Even with delayed choice, going backward in time, we always have a linear succession of events that went before.

Some people get hung up on parallel universes. This kind of question is not really scientific, because we can never verify the answer. By definition, parallel universes cannot communicate with each other. What amazes me is that the fundamental nature of reality rests in the impermanence of the world and the impermanence of the images that we make. Newton gave us a false notion of permanence and determinism—one antithetical to spiritual traditions like Zen and Hinduism. But now we are rediscovering the ephemeral nature of everything—the dynamic, creative, pulsating nature of everything. This is how quantum physics is integrating science and spirituality once again. When you heed the impermanence of the world, when you give up trivial pursuits and instead explore what really *is* permanent—the archetypes—you will have become a quantum activist.

CHAPTER 11

The Creative Principle

There is a tendency in the New Age movement to use the words "God's will" when discussing evolutionary change. In quantum science, however, we tend to use the concept of downward causation to explain evolutionary change and avoid more religiously and historically charged terms like "God's will." As I was developing the ideas presented here, I became more and more concerned about this phrase, "God's will." Scientists recognize that an irreversible principle like downward causation must be objective in order to be scientific—not objective in the Newtonian sense, perhaps, yet not dependent on arbitrary subjectivity. But creativity, in the quantum sense, isn't subjective. It has a subjective aspect that comes from individual human fallibility. It is not, however, the creative insight that is fallible; it's the mental representation of the insight that we make with our subjective minds.

God is the creative principle behind a creative insight. That much is easy to see, if we agree that what we call quantum consciousness is what religious people call God. God is then the perpetrator of downward causation. But what prompts a certain downward causation—God's will—to take place in a given situation? How does the manifest world come about? Carl Jung claimed that it was to make the unconscious conscious. That, as a general cause for any manifestation may be accurate, but it does not tell us why a particular unconscious possibility becomes conscious as opposed to another. In fact, it sounds very much like Darwin's biology—probabilistic, driven by chance. Evolution as described by Darwin can go this way or that way. But that is not how evolution actually works.

When we look at the fossil data, evolution appears to have directionality; there is order in evolution itself. In fact, this is one of its key characteristics. Fossils begin as very simple organisms and become more and more complex as evolution proceeds. Thus the passage of time is clear in the increasing complexity of the fossil record. But Darwinism doesn't have any directionality built in that favors complexity. Gene mutations are random, and random means without direction. Nor does nature select on the basis of complexity. It selects on the basis of fecundity—which mutation produces the most progeny. So Darwinism, as a theory of evolution, can never demonstrate an "arrow of time" in the direction of increasing complexity.

Quantum Evolution

This is where quantum science can help. In a quantum worldview, evolution is looked on as an evolution of the *representations* of the subtle possibilities of consciousness in the world. Among the four kinds of experiences—sensing, feeling, thinking, and intuiting—there is a gradation. Sensing is gross; feeling is subtle; thinking is more subtle; and intuiting is very subtle. In the context of evolution, there are representations of the least of the subtle—representations of the vital energies in matter that we call life. And more and more sophisticated representations of the vital are made as the evolution of life progresses. Organs become more and more sophisticated and associated with higher and higher chakras as time passes, as fossils increase in complexity. Thus the idea of evolution is to make the representations of the subtle possibilities of consciousness in the world *better and better*, as well as to make representations of the more and more subtle.

It was only after the vital-energy representations in matter were more or less completed that evolution produced the capacity for mental representations. With the development of the neocortex, the capacity to map the mind was introduced and evolution progressed with the improvement of the meaning-giving capacity of the mind. Are we humans at the top of the evolutionary totem pole? We could be, yes. In quantum theory, evolution does not regress; we do not need to worry about a planet of the apes in the future.

Mind first gave meaning to the physical world, so humans, when we were gatherers and hunters, were first to convey meaning to the physical world. In the next stage of evolution—at the time of small-scale garden agriculture—the mind started giving meaning to the energies of the vital world. Men and women worked together as cultivators. They interacted and had emotional battles and relationships out of which evolved the negative emotional brain circuits. Nonlocal tribal consciousness developed, so these interactions benefited all members of the tribe, even all humanity. Our ancestors may even have explored the archetypes to some extent, making representations that today we call Jungian archetypes of the collective unconscious.

Then came the age of mind giving meaning to mind itself—the age of abstract rational thinking. This began with large-scale agriculture. Landowners aggregated individual land holdings and tribal consciousness faded, along with nonlocality. Landowners rose to the top of a hierarchy; they had the power—they became the aristocracy and the rest became serfs. Rational thinking produces its own hierarchy. In the past, it produced a religious hierarchy that interacted well with the aristocracy. Even today, we struggle against this elitism.

The next step in the evolution of the mind occurred when the mind gave meaning to intuition and embodied that meaning through creativity. We can call this the intuitive or archetypal mind. This is why inner fundamental creativity is needed so much today. It also emphasizes the importance of integrating rational thought with emotions and feelings.

So what is the future of evolution after the age of the intuitive mind is over? As the process plays out, some day we will learn to make representations of the supramental directly in the body. We will learn to represent an archetype, to embody an archetype. In fact, when this occurs, we will be incapable of *not* embodying that archetype. Selectivity and preference will fall away in favor of the unitive, the collective. And this raises the possibility that this progression may continue, because, as soon as we have the capacity to represent the supramental in ourselves, we will ask the question: Are there higher experiences even more subtle than supramental archetypes?

I once had an experience that, in the absence of any other words to describe it, I will call satori or samadhi. The experience lasted for two days, during which my capacity for loving was effortless. I loved everyone and everything without making any effort. After two days, the ability gradually faded. But that is the closest I have ever come to embodying an archetype like love. For two days, it seemed that *I was love*.

I arrived at this experience through a meditation practice called *Japa* in Sanskrit, which I describe in my book *Quantum Creativity* (2014). I performed this practice very intensely for seven whole days. I think that the intensity was important. It was also important that I did the practice within the parameters of my regular lifestyle—my job, my teaching, my interactions with students, my interactions with my wife. In retrospect, I can see that there was an actual creative process present in the practice—a process whose effect of a loving "instinct," albeit temporary, I have so far been unable to reproduce. Usually, when we have an experience, we create the memory and we can relive that memory, at least in part. In this case, however, I have the memory of this experience, but I cannot bring back its effect. This is unlike every other experience I have ever had. Usually, I can actually "re-feel" the feelings involved in an experience and bring back some of the effect that accompanied it.

This leads me to believe that this kind of archetypal embodiment is a very transient experience, perhaps tied to a temporarily induced coherence in the action of the neuronal configurations. Whatever coherent configurations of the brain existed within that time period for me are gone and cannot be recreated solely through memory and will. So, to achieve the embodiment of the supramental, we have to develop entirely new capacities, perhaps entirely new coherent brain structures. Unfortunately, we have a long way to go. We haven't even finished the fourth stage of mental evolution—the intuitive mind.

Frankly, we don't know what happens beyond the embodiment of the supramental. We are creatures of the mind, and we can experience the supramental only as intuition—as spontaneous experience. But we don't have the capacity to represent the intuition directly itself as memory without the intermediary of thought and feeling. Only when we develop the capacity to render a direct representation as a more permanent memory will we be able to attain the next level of subtle experience. For vital beings like fish, which do not have minds, can mind be "intuited?" No, because fish don't even have an emotional brain. And early mammals have no idea of supramental values—they don't have ethics. We have to go all the way up the evolutionary scale to gorillas or chimpanzees to find ethical behavior. So I think we have to consider the next level of evolution before we can begin to see what comes after the archetypal.

Directionality and God's Will

The law that declares itself very clearly when we study evolution in the quantum frame is this: Evolution proceeds in the direction of making better and better brain representations of the archetypes through the intermediary of the mind and the vital body at this stage of evolution. The general statement of the law is that evolution proceeds in order to make better and better representations of the subtle. At our current stage of evolution, the subtle that remains to be represented is the archetypes. So evolution moves in the dirction of making better and better mental and vital representations of the archetypes, and better and better embodiments of them.

In a religious or New Age context, this is "God's will." In other words, God's will is purposive, and its purpose at this time is to make better and better representations of the archetypes—more love, more beauty, more justice, more goodness, more abundance, more wholeness. In a quantum context, we could say that God's will seeks out the purposive direction in which our consciousness evolves. We could say that God's will is exerted with the purpose of making the representation, the embodiment of the archetypes in us, better and richer, more and more profound.

But this purposefulness does not always manifest in the world. Surely, we can say objectively that, today, we are less violent than we were seventy-five years ago when we were fighting a world war. Today we can't even comprehend returning to a world in which war is the chief reality. Seventy-five years ago, provocative events like those occurring in the Middle East and in Eastern Europe today would have precipitated a world war. Today, we think through the consequences of such an outcome and allow our rational thoughts of economic necessity to overrule our instinctual tendency for violence and domination. The same is true of the free market, whose operation cannot be due to material interactions among elementary particles. The calculations involved in these societal actions and reactions are meaningful and purposive. So they must have a nonmaterial causal source--downward causation. And this can be expressed as the evolutionary movement of God's will in the world. The evolutionary movements of consciousness are the invisible hands that are moving these events.

Is there any reason to presuppose that God's will is different from our will as human beings? No. In creativity, we synchronize with God's will. When we manifest a creative idea, however, we sometimes cannot see God's will in every step, because we do not have the total picture. This can result in evil—as in the case of the atomic bomb. The mechanism for manifesting God's will through us is still faulty, so evil can occur even under God's will. This causes confusion. This mechanism will become better-oriented toward good as we evolve, however. Our representations will get better, and we will begin to get a better picture of what is compatible with good and what is not. This is how we will learn to avoid evil.

God's will is not self-organizing of God's intention. God—that is, the creative agent of the quantum consciousness—is not the self. Remember, self is created only with manifestation. Without manifestation there is no self. There is only consciousness and its potentiality for choice. All we can say is that this choice is objective—this choice, when collapsed, must be beneficial to evolution. That's the only criterion. If the choice produces better and better representations of the subtle, it manifests. It's as simple as that.

God's will is the causal power of God consciousness. Not *God's consciousness*, which sounds as if God were an individual, but *God consciousness*—nonlocal consciousness. God's will is thus totally objective. In the old religious worldview, God consciousness was somewhere distant from us. We prayed to it: "Please do such and such." If we think that God exists separate from our consciousness, we go back to the old tradition. But we know experientially that the quantum God may *appear* to be separate from us, but is not—although the ego does not recognize this. Every important evolutionary change takes place through manifest beings only. Remember quantum measurement theory: No event ever takes place without an observer. Even evolutionary movements are manifested eventually through manifest individual movements, albeit acting collectively.

The Ego and God

So are we separate from God in all events? No, we are not separate at the creative moment. When the appearance of separateness disappears, we become creative. In the experience of the quantum self, we see ourselves as separate from God only by the thinnest of margins—like Michaelangelo's Adam reaching out to touch the hand of God.

The separateness that we impose on ourselves is an imposition of ignorance that appears at the ego level and becomes further entrenched when the ego acquires a persona. And this ignorance is compulsory. There is no way out of it. The tangled hierarchy of the brain, along with the conditioning that results from the memory feedback of the brain, makes this separateness compulsory. Whenever we are experiencing, we experience separateness. The only exceptions occur when we experience those little moments of quantum movement—quantum leaps. It is possible to escape from separateness by entering into a sleeplike or unconscious state (I talk about that later). But in the waking state, the best we can do is to construct our lives in such a way as to become non-attached and avoid engaging our ego patterns. To escape, we have to live in such a way that we are not likely to fall into negative emotional brain circuits and into negative ego habits. It's tricky and difficult, but doable.

In fact, the question of whether God is separate has no meaning. Separateness is an illusion, an appearance. There is only one consciousness. We think that we are separate because of tangled hierarchy and because of the memories in our brains. The experience of separateness in human manifestation is, however, compulsory and the only way to escape it is through a quantum leap or two or to live in the unconscious.

My personal views on this are grounded in the worldview of quantum science. The God of quantum science is objective, scientific. God is the creative agency of quantum consciousness, which is nonlocal. God makes choices on the basis of a strictly objective criterion—will they move evolution forward or not? This can sometimes lead to choices that, in the short run, result in events that can be seen as evil. But God allows evil only because it may be needed for the long-term evolution of the species.

God also plays a second role in our lives, one we cannot deny. In moments of despair, we often call out to God in a very personal way. So what is this personal God? Remember that the self exists on two levels—the self that we call the ego, the conditioned self, and the primary self with which we engage when we are creative, the quantum self. The primary self is universal, but it manifests in an individual body-mind. We therefore have a personal relationship with the primary self, because we encounter it when we are creative. Creativity researchers have called this encounter *the flow experience*. When we have a flow experience, we flow smoothly between the ego consciousness and the quantum self-consciousness.

We know this self personally and, in moments of despair, we reach out to this flow. We reach out to the quantum self and wonder why it has conveniently become separate, why it doesn't come immediately to relieve whatever crisis we are facing. We cry out; we pray. And the quantum self does respond eventually. When this happens, intuitions begin to come. These are the experiences we focus on. But as a scientist, I know that the quantum self does not have any causal efficacy. That resides in the quantum consciousness or its agent of causation—God. Although no personal relationship is possible with this agent of causation—God, or consciousness—it is where all quantum causality resides. And I can only pursue that causality by aligning myself with the evolutionary purpose of humankind.

CHAPTER 12

Quantum Reincarnation

The concept of karma is central to a belief in reincarnation. From a Western perspective, karma operates as a more or less objective principle through which actions in this life determine outcomes in future lives. We accumulate karma through our actions and experience; we die; we are reincarnated in a way consistent with the karma we have accumulated. From an Eastern perspective, karma operates as a spiritual principle in which both intent and action influence future existences. We accumulate karma through intentions that contribute to either happiness or suffering in a karmic continuum; we die; we reincarnate as part of that continuum.

From the perspective of quantum science, however, karma is nothing but conditioning from past lives. A participant at a conference once asked me: "What about someone in poverty in West Africa who's living in a state of starvation, deprivation, and filth. They may sit all the time and intend that they're going to have money, but the money doesn't come. Does karma have anything to do with why money doesn't come to these poor folks?" In other words, he was asking the question from a Western perspective. He was questioning whether quantum science suggests that we are placed in this life for the God consciousness to manifest certain experiences in the physical body, and no matter what we do, that is the intent of *this* life in *this* body.

I answered him from a quantum perspective. I told him that I don't think our lives on earth are about suffering predestined karma; I don't think that death is just a transition from one predestined karma to another. I think karma is nothing but conditioning from past lives, and some kinds of conditioning can keep us from becoming expanded in consciousness. These we call *bad karma*. Conditioning is what produces the ego identity. Ego identity is a confluence of all of our conditioned habit patterns from this life *plus* some from past lives. Karma is thus just a big part of ego conditioning.

Indeed, bad karma may simply be conditioning that keeps us from manifesting creativity, from manifesting intention. By the same token, good karma may simply be conditioning that helps us be creative and fulfill our intentions. And most important, karma—like conditioning—is not compulsory. It is just a tendency. We can, in spite of bad karma and conditioning, be creative, and we can therefore expand our consciousness and overcome barriers to our intentions coming true. To do this, however, we have to know about creativity. We have to believe and engage in the creative process. The literature on this topic is rich. So if you are stuck in bad karma and conditioning, I say: "Wake up! Help is available." Explore some of the ideas I offer in my book *Quantum Creativity* (2014), which talks about the new science of manifestation as an aspect of quantum science.

Karmic Waves of Quantum Possibility

Believing in karma is tantamount to believing in reincarnation. It is difficult for some, however, to think that we literally die, get buried in the ground, and come back in another body.

When they try to explain how a material "soul" survives a dying body and reincarnates in another body, they get into trouble. Isn't it easier to believe that we, in all our incarnations, exist simultaneously, all at one time? In this model, a reincarnational memory actually taps into other lifetimes as we experience them simultaneously. In a quantum worldview of potentiality and nonlocal access, this makes sense, but it is not a model that agrees will all the empirical data.

In quantum science, the essence of what we call the soul consists of habit patterns—mental and vital—and that essence continues even after death. So we can experience these continuations, these continuing patterns, in a future time and place in a future body. This is a valid interpretation of what happens in reincarnation, and data actually supports it. In this view, the soul is not made of material substance—not energy, not subtle substance, not even subtle energy. It's not substantial in any way. It is simply a pattern of habits. I call these patterns *nonlocal memory.* If someone in the future uses (or inherits) my memories, that person can legitimately be called a reincarnation of myself—but not in the sense of a substantial soul traveling from my body into a future body. Transmigration is more subtle than that; that subtlety has to be understood if you want to accept reincarnation in a quantum frame. In my book *Physics of the Soul* (2001), I offer a complete explanation of this and show how quantum science can be totally compatible with reincarnation. Moreover, the data is there to support this explanation.

As long as you recognize that the physical body is gone forever at death, you won't go too far wrong. Some of our burial rituals can be very misleading in this respect, because we tend to think that something is preserved in the buried body. No, the physical body is simply dead. It's finished. But the more subtle aspects of ourselves—how we use our vital, mental, and supramental bodies, our consciousness itself—live on. The patterns of how we use the experiences accumulated during our lives—the patterns that define our ego character—live on in some very real sense. And these patterns can be recycled. And that recycling is what we call reincarnation.

Other researchers continue to think in other ways, however. For example, the consciousness of the soul is often described as a frequency or a field. But then questions about how this frequency or field is contained arise, causing further confusion. This kind of argument uses language common in science before quantum theory developed—it approaches karma from a Western perspective. But if you change the language of this argument just a little bit, it catches the essential truth of what quantum theory is saying. In other words, don't take the word "frequency" literally. Think of it, rather, as referring to a wave-ness—a quantum possibility wave. The essence of the soul as I describe it—these patterns of nonlocal memory that reincarnate—are changes in the probabilities of quantum potentiality, changes in the probability associated with quantum possibility waves. So they resemble the modulation of a frequency, as in an FM radio. They are somewhat like frequency modulations—not the frequency itself, but the modulation of frequency. And we experience a similar modulation of these waves of possibility as we live our lives. In other words, the probabilities of these possibilities are modulated by how we experience life, by how we condition ourselves, and by how we develop our patterns.

Although this is not unlike how frequency-modulated waves go from one region to another, it is not exactly like that. You have to understand the analogy, not in terms of something like a wave or energy actually traveling a finite distance as a signal. Instead, you have to understand it in terms of patterns of usage of potentiality that are being transferred nonlocally from one place and time to another place and time.

The really big question is, why? Why go through this melodramatic cycle of birth, death, and rebirth? That is a question for a later chapter.

The Sleep of Death

From an Eastern perspective, when someone dies, people say that person "passed away to the other side of the world." Indeed, sleep is seen as equivalent to a small version of death, and waking up as a small version of reincarnation. We fall asleep; we enter a spiritual state (the unconscious); we awaken.

From a quantum perspective, the experience of sleep is a good place to start when discussing the nature of death. In quantum physics, sleep is the state of the unconscious, where no subject-object split exists. Sleep is necessary because the mechanism of tangled hierarchy in the neocortex requires rest. Similarly, we know that the overall mechanism that we have in the brain for making and retrieving representations of subtle meaning becomes atrophied as we grow older unless we engage in creative activities. Indeed, this becomes more and more evident as we approach middle age. When we reach old age, we encounter physical and mental problems that may seem insurmountable. So in the quantum worldview, it seems right that death is like a long sleep.

In fact, the whole mind-body mechanism needs to be renewed. And this, in a sense, is the essence of reincarnation. But when we sleep, we wake up fully cognizant of continuity with our previous waking states. How can we ever know that this continuity survives after death—how can we know that at least a part of us survives in a future brain and body? Babies, for instance, don't usually share reincarnational memory of adult language. They have to learn to speak from scratch. Yet children all over the world do remember past incarnations and share them. So it becomes very tempting to assume that at least a part of us not only survives death, but also reincarnates. And quantum physics can give a verifiable positive solution to this paradox of survival and reincarnation.

Let's continue the analogy with sleep. How do you know when you wake up that you are the same person? You know because there is a continuity of memory. Is there an analogous continuity of memory in some way from one incarnation to another that applies to all people? Our tendency is to believe that all memory resides in the brain. When we apply quantum physics to the brain and the correlated mind, however, we get a concept called quantum memory—memory that resides outside of space and time, and not in the brain at all. So with quantum theory to guide us, survival after death and reincarnation become quite plausible and feasible.

Character—The Continuity Gene

Some traditions see the continuity necessary for reincarnation in another way. For example, in one life, two people who have a troubled relationship end up, in their next incarnation, as husband and wife in order to overcome the challenges that thwarted them last time around. This implies strong continuity of memory. But this is another kind of continuity, different from the continuity of quantum science. This is continuity of the storyline itself, between people who stay nonlocally correlated across incarnations. The concept of soul mate uses a similar logic.

It is crucial to remember, however, that what defines the ego is only partly the history that we create. There are things besides the storyline that define our ego identity. In fact, the storyline is actually created as a side-effect of what is really important—namely what we learn, or the pattern of learning itself, the pattern of habits that we create around our learning. The storylines of our lives often undergo major changes. Some may discover that their storylines really didn't matter. Very few will defend their storylines—like a rich person who becomes poor and vice versa. Most people take this philosophically; they recognize that character essentially remains the same, irrespective of storyline. Even in one incarnation, we must give more weight to character, which, once formed, does not change, even though the storyline may keep changing.

But if quantum memory applies to character, and character is stored nonlocally, then there must be some indication that we value our character more than our storyline. When we look at ourselves, we generally find that our storylines are not all that different from other people's storylines. A Sufi parable tells of a king who asked his wise men—his magi—for the essential storyline of human life in a one-sentence summary. They all agreed on this sentence: We are born; we suffer; we die. Who can object? Although I prefer something like this: We are born; we suffer and learn; we use suffering as a springboard to explore creativity; we die. But whether we suffer fruitlessly, or whether we use suffering as a springboard for creativity or for more suffering, is an issue decided fundamentally by our character. So the quality of your life depends entirely on your character, although your storyline may be essentially the same as that of another person.

In Japanese culture, there is a type of character called *samurai* that depends entirely on a warrior's defense of his honor. Consider this story:

> A samurai's parents are murdered by an enemy. Social pressures dictate that the samurai must defend the honor of his family by fighting and destroying that enemy. The samurai goes to battle and fights with great courage. In the end, he defeats his enemy and is about to kill him when his enemy looks at him with great hatred and spits at him. To everyone's surprise, the samurai walks away.
>
> His relatives and companions are very angry. "You let him live to raise another army," they cry. "What explains your strange behavior?" The samurai replies: "When he spat at me, I momentarily felt a great hatred. As a samurai, I vowed never to kill for personal reaons. But hate is very personal. I could not kill in a state of hatred. My personal honor must take precedence over my social and family honor."

Many ancient cultures had a similar concept of honor. Julius Caesar crossed the Rubicon to defend his honor. King Arthur's round-table knights were famous for defending their own honor, as well as that of others under their protection. What the quantum worldview proposes is that this character—the pattern of habits, the pattern of learning that each of us lives—is nonlocal. Character is not stored in the brain, but exists outside of space and time. And it is the character that is reborn from one incarnation to the next. Notice also that this ongoing character is dynamic; each incarnation contributes to it. Many people call this surviving entity the soul. In my book *Physics of the Soul* (2001), I call it the quantum monad. In the world of possibility, of potentiality, many possibilities reside together in the One.

But this raises an intriguing question: What triggers a birth event? What triggers the choice of the father and mother to join and conceive and bear a child? From the point of view of the quantum theory of reincarnation, we can say that the child has a certain predisposition that exists, even before it is born, to be part of a particular family with particular parents. The character, which is nonlocal memory, has to find an abode in a new environment, in a new body, in a new brain. And the construction of this new brain and body depends on the genes that are contributed by the parents. The make-up and development of the child will depend on the makeup of the family and on cultural norms. In other words, attributes of the form depend on the genes; attributes of development depend on the culture and the family.

I suggest that there is a universal law of reincarnation by which a coordination takes place between the various potentialities and propensities—the pattern from previous incarnations or history that the child expresses, the genes that make up the physical body, the culture into which the child is born, and the upbringing the child is likely to have after he or she is born. In ancient Greece, Plato talked about an individual's soul (the quantum monad) choosing its parents. But that choice could be a part of a universal law. In other words, the universe could conspire in such a way as to create a synchrony or congruence between the propensities born and the genetic and cultural makeup into which those propensities are born.

A child is thus born with certain physical characteristics represented by the genes and with certain predispositions. And that child will grow in an environment that will influence its conditioning in early childhood. But that child is also born with certain predisposed propensities that are available to it from past incarnations. These are just predispositions, however. In order for them to actualize, the child's environment has to trigger their use. Sometimes there is a great mismatch in early development that prevents the child's inherited elements of character from past incarnations from finding proper expression. For example, a so-called "idiot savant" may be exceptional in a specific area, but may be mentally handicapped in most other aspects of life. The child's upbringing is such that his or her character remains developmentally delayed; it never has the opportunity to manifest or to trigger the genius that it brought from past incarnations. If, all of a sudden, this individual experiences a trigger, those genius propensities find expression. We quite frequently find this phenomenon in autistic children.

The Law of Inherited Karma

So why does a particular child inherit a particular bunch of past-life propensities—a particular character? Why not others? This is a question in whose answer lies a new law of the universe. From quantum physics, we know the concept of nonlocal memory. But the empirical data suggests that no two people use the same confluence of reincarnational propensities; reincarnational propensity rather travels only from one single being to another single being in a linear chain not unlike a string of pearls. So we postulate that there is a law of the universe (apart from the physical laws) that dictates a complete chain of people who are correlated, entangled together, who will participate in these successive incarnations of a more general existence that I call the quantum monad. This ongoing, developing nonlocal character, which occurs outside of space and time and beyond separate individual incarnations, belongs to the domain of potentiality. So, in an ongoing dynamic way, all of these reincarnations use the propensities contained in the quantum monad (see Figure 14).

In the figure, the first character (on the left) is an incarnation of a quantum monad that resides in oneness. The second character (in the middle) is a *re*incarnation of that representation. The third character (on the right) is a subsequent incarnation of that reincarnation. All three exist in potentiality, waiting to be born. When the first dies, there is another incarnation that, in the future and by universal design, will use the ongoing propensities of that previous life. When this second incarnation dies, a third will appear. This process continues according to a law of the universe that I am proposing is a kind of a law of inherited karma. And the synchrony of the parents, and the culture, and the upbringing, along with the propensities that are available to the reborn baby are all part of this law—the law of inherited karma.

Let's suppose that I am the first person in this series. I am born; I live a storyline; I die. My storyline is not reborn as memory in my new incarnation's brain. But behind that storyline, there is also character. The storyline is the content of my existence—my name and other details. For example, I was born in a small town in India near Calcutta. That content also includes specific sufferings, like the malaria I had in childhood or the trauma of a riot I witnessed between Hindus and Muslims. But the brains of newborn babies are not loaded with storyline memories of past incarnations. My current storyline is all that is in my brain memory and that memory will die with me. It is true that, as quantum possibilities, this memory is forever in the domain of potentiality, but without the brain it will not be reinforced again and again by recall. So it will be harder and

Figure 14. Propensities (karma) creating an ongoing string of pearls through birth, life, death, and rebirth.

harder for future incarnations to retrieve it. Even so, a surprising number of people, especially children whose identity with their past lives is still strong, do recall past-life memories, as psychologist Ian Stevenson's data suggests.

Now let's return to the character. How I dealt with my storyline as it was developing in my childhood, how my trauma challenged my upbringing and catapulted me somehow into the field of physics—all that is part of my character. The experiences I had in constructing my storyline established patterns of behavior that contribute to my character.

Empirical data of near-death experiences suggest that, at the moment of death, we are given a panoramic experience of our past, present, and future lives. This is our last chance to influence what is going to happen in the future, which is still potentiality. This panoramic experience at death is part of our cognitive experience. And we have a chance of influencing the future through what the new quantum science calls intentions. In a way, we each can attempt to create an important part of our future incarnations through our intentions during this panoramic experience. If our intentions resonate with the intention of the totality, then they may even come true. But does that mean that my storyline continues? No.

You really have to be dispassionate about this. Is the baby who is going to inherit the propensities that you created, *you* in some way? Not the whole you, certainly, but an important part of you, because you contributed to the character of this future person. In other words, we substantially contribute to the future, and that is very good to know. On the other hand, it is wise to give up the idea that you—as the hero of this storyline, of a lived life—come back to continue the storyline. And in fact, would you really want this to be so? If you think about it, by the time you die (if you die of old age), your storyline has played out. It's time to move on. The *you* in the new incarnation will be a clean slate—a fresh start. No one wants to be born carrying the burden of a tired storyline. This is how quantum physics can help us escape the rigid determining laws that govern life and death in a Newtonian worldview.

Near-Death and Near-Birth Experiences

The panoramic view experienced at the moment of death represents a nonlocal view of a person's past, present, and future. But don't think of that future as concrete. The future is always potentiality. You can influence the probabilities of the future with your intentions. And those intentions can certainly generate a highly probable possibility for influencing your actual rebirth. What we can learn during life is how to make intentions that resonate with the intention of the whole. For instance, imagine a dying man who has a specific vision of being reborn as a child with certain parents into an environment that would be perfect to express his intended character and impact the world maximally. If his vision resonates with the totality of potentiality, with the movement of consciousness, this child will then be born in actuality.

There are many reports of things like this happening. But these are details about which the theory does not have much to say—only the very general comment that, yes, it is possible. With science, all we can verify is that the death experience is nonlocal, as these experiences suggest.

Is the death experience a nonlocal opening, as these panoramic visions seem to indicate? The evidence suggests that it is. There is substantial data suggesting that, at the moment of death, a feeling of intense joy is experienced by people surrounding a dying body. In fact, I think it is probably fair to say that one of the reasons some people volunteer as hospice workers is to experience this deathbed, nonlocal experience of joy and peace. We also hear from people who have had near-death experiences that, as they left their bodies, they were surrounded by an unseen presence and felt quite blissful. And this nonlocal connection goes both ways. Many people have reported seeing the vague presence of entities around a deathbed. And indeed, some near-death experiencers describe meeting their relatives and spiritual teachers, images that are easily recognized as Jungian archetypal memories from the collective unconscious.

There is also objective evidence that near-death survivors have nonlocal experiences. For example, people describe autoscopic visions in which they looked down on their bodies on the operating table from the ceiling. They have even described details of the surgery that revived their brains. This is convincing empirical proof that nonlocality is indeed possible at the moment of death. And it does happen. When we die and our identities move into our subtle bodies, consciousness stops collapsing possibilities into actuality because the brain is dead. So if a person is revived, all these experiences that near-death patients describe must have happened retroactively through delayed choice—a delayed collapse.

These near-death experiences also give great support to arguments for the phenomenon of reincarnation. Clearly, they show that survival is a fact after death, because otherwise how could patients remember anything that could be said to have "happened" after death. Although they are recalling only a *memory* of what happened, the events they describe still happened in some real sense.

Likewise, at the moment of birth, there may be some recall of events from delayed choice; the brain may have made memories of those recollections. Of course, a newborn baby cannot share its birth experiences. But in some of Stanislov Grof's experiments with people doing a special kind of deep breathing called *holotropic* breathing, subjects have regressed to the time of their birth and remembered the experience of going through the birth canal. They also remember things that could only have been experienced retroactively preceding the birth.

So when I die in actual three-dimensional reality, my body dies and my "I" disappears. But we need to be very clear here about what the "I" is. Remember, any given existence has two components. One is called the storyline—the content of an individual's brain generated by experiences. But on a deeper level, what defines that existence is the character, which processes things in a particular way. For instance, a person who explores the archetype of love through mathematics takes on the character of a mathemetician. A person who explores the archetype of beauty takes on the character of an artist. These are character elements. And quantum science tells us that character is a deeper aspect of the ego than personality or storyline. It is the part of the ego that is stored nonlocally, as nonlocal memory.

Content vs. Character

Neurophysiologist Karl Lashley devised an experiment that suggests the distinction of content from character. Lashley was trying to find the location of memory in the brain that is activated when we learn a behavior. He taught rats how to find cheese in a Y-shaped maze. One branch of the Y contained cheese; the other branch gave an electric shock. Of course, the rats learned very quickly how to find the cheese. In order to find where the memory of that learning was located, Lashley began removing parts of the rats' brains, assuming that, if the memory—the learning—disappeared, it must have been contained in the portion of the brain removed.

Even after removing 5 percent of the brain from five different locations in five different rats, however, all the rats could still find the cheese. So he repeated the process removing 10 percent, then 20 percent. But even after removing 50 percent of the rats' brains—even though they could no longer see or walk—the rats still crawled through the maze and found the cheese. Lashley concluded that the memory of learning must reside throughout the brain. Neurophysiologist Karl Pribram, Lashley's student, developed a holographic theory of the brain based on this conclusion that enjoyed popularity for a time.

Of course, an equally valid conclusion would be that learning does not reside in the brain at all. And this is the view of quantum science. Quantum memory, nonlocal memories, do not reside in the physical brain. Brain memory consists of individual episodes of learning—events that construct content—but the propensity learned transcends brain memory and is stored nonlocally. The individual episodes of learning that took place when the rats found the cheese were stored in the brain, but the overall propensity that resulted from all those individual episodes of learning were stored outside of space and time.

In the same way, my propensity to think like a philosopher-scientist is stored nonlocally, although all the individual experiences and stories that resulted in my learning to think this way are stored in my brain. I sat and read and pondered on what I read in a certain way at a certain place at a certain time. If that episode was significant, it was stored in my brain so that it could be recalled. But my entire habit—my character, my propensity to philosophize in a scientific way—is stored outside of space and time and can be inherited by the baby who is my next reincarnation. That child will thus have a ready-made predisposed propensity for philosophizing in a scientific way. If that propensity is triggered, the child will be called a "born scientific philosopher" by his peers.

Suppose I spent my early years as a self-centered child. As I matured, however, I became other-centered, nicer. Toward the end of my life, I might even be regarded as "full of love" and "loving." But what is my character? There are two operative factors here: I learned to change, and I transformed my self-centeredness and learned how to love. That *how to change* and *how to love* will then be elements of my character and be stored in the nonlocal memory. In the next incarnation, when these propensities are needed and are triggered, they will be recalled and will become manifest in that child. That is the way it works.

The question of triggering propensities is crucial here. The transfer, or transmigration, of character requires that the reincarnated child be born into an environment where such triggers are readily available. Suppose you are born into an affluent life and go through the change from nastiness to loving. When you die, that memory is stored in the nonlocal memory. When you are born again, it does not matter if you are born into wealth or poverty. You will still be loving in relationships once that loving nature is triggered. Love is the easiest archetype to trigger in any life situation. Or consider the case of a rich man who gradually evolves from being a miser to being generous. In the next incarnation, he is born once again as a rich person. In this case, triggering is easy, because the opportunities for generosity are enormous. If he is born into a poor family, however, generosity will not be as easy to trigger.

Take the case of French mathematician Évariste Galois. Galois was born into a very non-mathematical family. He was not exposed to mathematics in any way until he went to school.

One day, he accidentally came upon a book on geometry and read it. That event became a trigger for him to become a great mathematician with very wonderful ability. This kind of "crystalizing" experience, as it is called in the creativity literature, is an example of a trigger (see John Briggs' 1990 book *Fire in the Crucible*). And perhaps that trigger was activating a propensity—a habit of character—brought forward from a previous incarnation.

What is the trigger for someone to be born into this life, into this particular time and place? This is where the operation of a universal law that controls these triggers becomes essential. The triggers that precipitate a particular birth in a particular time are part of a universal law, which we have to decipher empirically. Remember, being born is not part of the propensities that are stored nonlocally. The individuals who are, who will be, part of that ongoing string of pearls whose developing locus of propensities we call the quantum monad, will be part of this *particular* string of pearls because of a universal law of correlation between these individual incarnations across time and space. Only these individuals will benefit from these particular ongoing, developing propensities. In that sense, all these immanent incarnations—the entire correlated string of pearls—can be called incarnations of that one quantum monad. And the time at which the quantum monad prompts a specific rebirth probably depends on the amount of learning, the amount of maturity involved. The less maturity, the more frequent the rebirth.

CHAPTER 13

The Meaning and Purpose of Life

Clients often ask me what the meaning and purpose of a human being here on Earth is. What are we doing here? And does the quantum worldview give us hints as to that purpose?

In simplest terms, the answer to these questions is that we are here to satisfy our souls. So what we should ask ourselves is: What really satisfies me? Pleasure may seem to satisfy for a little while, but the energies involved in this satisfaction actually reside in what we call the navel chakra, not in the crown chakra where real satisfaction—wholeness—is experienced. Thus the feeling of soul satisfaction never comes from pleasure-directed pursuits. Only when we feel deeply positive feelings associated with love or any other archetype do we actually satisfy the needs of the soul. The soul, in this context, means the supramental or archetypal body—or, more accurately, the mental representations of the archetypes that we make that some traditions call the "higher" mind.

When we explore meaning and the archetypal values that Plato talked about—love, beauty, justice, truth, goodness, abundance—we feel happy and satisfied. When we embody these values and archetypes, we make our souls richer in content. This is the goal of our evolution. We can pursue this goal in our lives by applying this very simple criterion to our thoughts and actions: Is the act that I'm about to perform, is the experience I'm about to choose, going to take me toward wholeness or away from it? When we use this simple test, we become aligned to the movement of consciousness. Quantum activism is just another step in this process.

The Power of Crisis

It sounds simplistic, but it is true. And if true, why haven't we become more aware and conscious of this simple way of living? Is it just a matter of publicity, of letting the truth be known? Wouldn't the world be better for it if everyone followed this simple path? Find satisfaction, and become aligned with the movement of consciousness.

Well, in practical terms, of course, we generally need some sort of crisis to wake us up to this truth. Without either personal or social crises, we tend not to bother with change. For some of us, a personal crisis of confidence or a personal crisis of unhappiness makes the difference. Once when I was at a conference, I spent a whole day being jealous. At the end of the evening, a thought came to me, and that thought changed my life: Why do I live this way? Why was I letting lack of integration and separateness direct my life? Why was I choosing a path that I knew would leave me dissatisfied—in this case, the science I was pursing? Of course, nobody *told* me the answer to these questions; the answer came only after many years of exploration. It takes a little bit of tenacity. But when you experience a crisis like this, it gives you the tenacity that you need.

For some, a personal crisis isn't enough. They need to experience a social crisis. Well, look around you. We are in social crisis now, with global climate change and terrorism and the breakdown of the economy. We are experiencing a breakdown of democracy, as well as a breakdown of our healthcare system. But when we become aware of these crises, and of the shortcomings of the current unsustainable materialist consumer-driven economy, how can we change it?

I think many people—especially business people—will become quantum activists in the coming decades because they are already feeling the crunch. They know that we cannot just go on running the economy the way we have been—without innovation, without developing new sectors into which the economy can expand, without paying attention to social good and sustainability. The information industry may still be going strong; people are still buying new generations of cell phones. But how long can this continue? There is no meaning and satisfaction to be found on our current path. Consumerism and the materialism that drives it are no longer forces that satisfy. In the meantime, we are losing our jobs to robots—jobs that, in turn, gave us money to engage in consumerism.

So what can we do? As a scientist in pursuit of integration, I suggest extending the economic arena to include spiritual commodities that can rescue consumerism and create meaningful jobs to replace those lost to machines. In my book *Quantum Economics* (2015), I suggest that we bring the production and sale of subtle and spiritual commodities into the economic arena. But how can spiritual commodities like happiness and love be bought and sold? How can they be produced *en masse*? I suggest this can happen when we change our attitudes. If you want to become a merchant of happiness instead of a merchant of violence, stop selling weapons and start selling love. That, I guarantee, will solve your personal crisis and make you a very satisfied person. That's the kind of change that will bring business people to a quantum worldview and make them quantum activists.

Poker Karma

Quantum physics presents a wonderful solution to the problem of the purpose of a particular life—nonlocal memory. We established in the last chapter that individual incarnations form an evolutionary continuum—a manifest string of incarnations of an ongoing confluence of quantum memories (karma) that we call the quantum monad. The source of our string of pearls.

But how does an individual pearl get the benefit of all the previous propensities, all the past karma that has been accumulated so far. There are so many propensities—so many conditioned memories—how can we focus on any particular exploration of an archetype? We can be attracted to too many projects and thus try to use all the multiple talents that have been triggered in us. As a result, none of our explorations may be forceful and achieve success. Satisfaction may elude us. Hindus have an answer for this. They theorize that there is a vast accumulation of karma at any given time, but that we don't bring all of it at birth. Only a selected part of the accumulated propensities are brought to a particular birth. And then, of course, we add more propensities to the total cumulative karma. That is called future karma. There are thus three kinds of karma: accumulated karma, indigent destined karma that is brought to a current life, and future karma that is created in this life.

Past-life regression therapist David Clines has had the opportunity to study a very large number of clients who go into past-life regression under his guidance. For many of them, he has recorded more than one incarnation. Clines was surprised when he realized that the propensities of a life and its subsequent incarnations did not demonstrate much continuity. The pattern he saw rather fits a hypothesis he called "poker karma."

In poker, each player begins with only five out of fifty-two cards. Clines theorized that people, in a similar way, do not bring forward all of the accumulated karma in a particular incarnation—a full deck, so to speak—only a portion of it. His empirical data and his conclusion support what the Hindus theorized millenia before—that we bring only a *part* of the accumulated karma to any particular incarnation.

But we must take the next part of Hindu theory seriously. Why do we choose a particular subset of karma—why those particular five cards? Hindus say that we bring those particular cards to the table because we want to fulfill a particular learning agenda in this life. Every incarnation has a purpose, which is to fulfill the learning agenda that the incarnation was meant to fulfill. Hindus call this idea *dharma*.

Of course, sometimes we go astray. Due to life circumstances, the needed propensities to pursue our dharma may never be triggered. Perhaps we don't pay attention to our talents, or we don't pursue the right meanings and archetypes of those talents. Suddenly, we are in crisis. We intuit that we have to change and, just as suddenly—through an event of synchronicity, a crystallization experience—we discover our learning agenda, our dharma. When we follow it up with conviction, it brings us satisfaction. If the propensities are never triggered—or if triggered, do not match what we explore—then we will remain unsatisfied. Many people live their lives without ever feeling satisfied.

How do we know what cards we have been dealt? Important question. If we are very fortunate, the crystallization experience takes place naturally. But if it doesn't happen in the natural course of life, what can we do about it? In my workshops, I take people through memory-recall experiences so they can remember their childhood propensities and discover the ones they have *not* been using. Perhaps there is an archetype that brought them satisfaction as a child, but they have not been pursuing it because of circumstances in their lives. In Grof's holotropic breathing exercise, some people spontaneously achieved a similar memory-recall experience through which they remembered the objective, the purpose, that was driving their current incarnation. Others may be able to recall their archetypal objective by exploring synchronicities and Jungian archetypes through Tarot. As Joseph Campbell used to say: "When you discover what you came here to do, you've found your bliss." And his advice was: "Follow your bliss."

I myself had a crystalizing experience that began my journey of integration at the conference I mentioned before. As a nuclear physicist, I was using quantum physics in my research, but I never bothered to understand the meaning of it. I approached physics in a very divided way, as a profession that had nothing to do with my life, with the way I lived. And this separation created a lot of unhappiness. Then I was forced to confront this unhappiness. I was invited to give a talk, which I did. But when the other speakers delivered their talks, I found that theirs were better than mine and that they were getting more attention. I felt extremely jealous and insecure. I remained immersed in that jealousy and negativity all day and into the evening. Finally, at one o'clock in the morning, I found myself with raging heartburn and an exhausted packet of antacids.

Feeling disgusted with myself, I went outside, where the cold air from Monterey Bay hit me in the face. Then a thought came: Why do I live this way? And, with that thought, I knew that I had to integrate my life and my work. So integration became my theme, the purpose of my life. And it has brought me satisfaction.

If you try to play poker without knowing what your five cards are, you inevitably lose. When we don't find out which cards life has dealt us, we go through life in turmoil. But when events take place, those experiences trigger in us a need to explore—a need to find out which cards we have in our hand—and suddenly we know the archetype we were meant to pursue.

For me, it was the revelation that integration, pursuing the archetype of wholeness, was the purpose of my life. And fortunately, one of the propensities I needed to follow my purpose had already been triggered in me—the scientist. The other, philosophizing, I had to learn in a hurry. For Galois, the crystalizing experience was the discovery that mathematics, pursuing the archetype of truth, was his purpose. But we are the lucky ones. For many, their characters don't match the professions they are pursuing. So they are unhappy and continue to question the meaning, the purpose, of their lives.

Archetypal Professions

The professions that we choose to follow are all directed toward exploring an archetype in the Platonic sense. Scientists follow the archetype of truth; artists follow the archetype of beauty; almost everybody follows the archetype of love; religious leaders follow the archetype of goodness; business people follow the archetype of abundance; healthcare professionals follow the archetype of wholeness. In fact, everyone has an archetype to follow. If the archetype of the profession you pursue matches the archetype you were intended to pursue, then you follow that archetype happily. You are happy in your profession.

In my case, there was a mismatch. I was pursuing traditional physics when I should have been following the archetype of wholeness. Most scientists today do what they do to make a comfortable living. They seek success, maybe even power, but they don't explore any archetype. So their professional lives are not satisfying, and they don't know why. When I learned to go deeper—when I started *really* exploring quantum physics from the standpoint of investigating wholeness rather than just truth, the scientist's usual archetype—I discovered an integrative worldview, and I became happy.

In the business world today, there are many people who only want to make money. But business is really an investigation, an exploration, of the archetype of abundance. When you engage in business from the point of view of abundance, you will find success in that exploration; you will become happy. Business people who are no longer following the archetype of abundance they set out to follow, or for whom abundance is not the archetype they were meant to follow, will be unhappy. A mismatch has taken place.

If you are unhappy in the pursuit of your profession, you have to discover what there is in the way you are following the archetype of that profession that is not fulfilling anymore. Are you in the wrong profession? In that case, change professions. Or have you not found the proper match of your propensities within your profession? In that case, change the way you approach your profession.

Sometimes a profession may not allow your particular propensities to be expressed because it has become very constrained. Then a thorough investigation of the archetype within the profession may have become almost impossible. In that case, the only recourse you have is to become independent, to go outside of the mainstream of the profession and devote your life to changing the nature of the profession itself. Become an activist. For me, scientific materialism became a straight jacket I was forced to wear because my profession demanded it. I had to break loose.

In general today, I find that many turn to activism because society has made it nearly impossible to explore the archetypes within the fixed professions that have evolved from tradition. In Japan especially, there is a tremendous mismatch between traditional values and the materialist culture that is overwhelming the country. But we see this in America as well. The professions have become narrowly defined, because the worldview that defines them is driven by scientific materialism. But the practitioners of these professions were born into families that lived according to the old worldview—that mind and matter are both real. The propensities they brought to those professions were thus meant to explore archetypes, not materialist values like money, sex, and power.

The Archetype of Love

Love is one of the major Platonic archetypes. Quantum physics sees the exploration of the archetype of love as an expression of the creative nature of that exploration. Archetypal love has two traditional aspects. One sees it in relation to the heart chakra. The other sees it in relation to its morphogenetic field, the blueprint of biological form and function in the thymus gland, which performs the function of distinguishing one body from the other. We can use quantum language to understand this association. The blueprint of form and function is correlated with the thymus gland because the thymus gland is built from the blueprint. So, the blueprint is part of the vital body, and the organ is part of the physical body. This vital-physical pair is what we must focus on.

This is the language that the new quantum psychology uses when it talks about the chakras. In quantum psychology, each chakra is a confluence of physical organs and their vital blueprints. In every chakra, we experience a particular feeling that is connected with the movement of that chakra's morphogenetic field—its vital blueprint. What we feel is vital energy—an energy moving in and out of that part of the vital body. It is not the movement of the physical organ, but rather of the vital energy of the blueprint correlated with the organ. In the case of the heart chakra, when the immune system (in the form of the thymus gland) denies a distinction between me (my body) and not me (what is foreign to my body), its function is suspended and I feel the possibility of love for another person. Suspension of the immune system makes me feel that I am one with this other person, that I am the same being as that person. So I get the feeling that you are mine and I am yours. And this feeling expresses itself with a strong intention of bodily union. And that is what we call romantic love.

Romantic love is one way in which the exploration of archetypal love can start. Romantic love is a very physically oriented love. But it still has a quantum aspect, a nonlocal correlation aspect—the correlation between vital feelings and the functioning of a physical organ. The bodily aspect of romantic love is eventually expressed in terms of "molecules of emotion" like the pleasure molecules called endorphins, because the brain comes into play whenever sexuality is involved. But these molecules only have a certain time during which they remain active. They do not remain active with the same intensity with the same partner forever. They eventually stop due to habituation. This is when romantic love runs out of gas.

Love that keeps the immune function suspended is important, however, because it gives that system a rest that is very important for it to function properly. Just as the neocortex gets to rest when we sleep, the immune system gets to rest when its function is suspended because we are engaged in loving someone.

But can love continue even when the molecules that mediate its physical pleasures fade? Yes, if we engage creatively with unconditional love. When you love someone even when the sexual energies driving that relationship are no longer very active, this is the beginning of the quest for unconditional love. As a beneficial side-effect to your quest, your immune system gets the rest it needs. This quest does not require sexuality. When you explore love creatively with unconditional love in mind, you delve into the creative quantum nature of love—quantum love.

Moreover, creativity is only one aspect of the nature of quantum love. New research tells us that the immune system has a sort of autonomy that is analogous to the autonomy found around the neocortex. This prompts us to ask whether there is a "self" associated with the heart chakra like that brought about by a tangled hierarchy. In other words, is there a self in the body? Is there an identity of consciousness associated with the heart-chakra organs? We know that consciousness identifies with the neocortex, with the brain. Can we also say that consciousness identifies with the heart-chakra organs? And yet, the immune system does not have a tangled hierarchy among its components. So needless to say, when I looked at this problem, I was very puzzled.

Then I looked at the problem again from the viewpoint of Traditional Chinese Medicine. In Chinese medicine, when they talk about organs influencing each other, they are not talking about physical organs, but about the vital correlates of physical organs that interact through pathways called meridians. For example, there is a meridian from the vital correlate of the liver—a navel-chakra organ—to the vital correlate of a heart-chakra organ. From a quantum viewpoint, this means that a tangled hierarchy exists between the vital blueprints of the navel-chakra organs and those of the heart-chakra organs—a tangled hierarchy that involves the vital body and two different chakras. This tangled hierarchy does not cause the compulsory self-reference that comes with the collapse of the possibility wave, as in the brain. That requires a tangled hierarchy in the physical body. But if we can identify, even at the vital level, with a tangled hierarchy, couldn't we develop the sense of a self—one that can be called the heart self or the heart center of the body?

Self-Love and Other-Love

Is the heart in this model the same as the heart talked about in spiritual traditions? I think it is. Moreover, it may even be possible to establish a physical tangled hierarchy associated with the vital one. This may be through the intermediary of the bioelectric body that can be measured with Kirlian photography that reveals an actual aura—a bioelectric body that we have in addition to the biochemical body of which the organs are part. And if this bioelectric body contributes a physical tangled hierarchy that complements the tangled hierarchy in the vital body—a hierarchy that we can cultivate and identify with—then we will be able to collapse feelings independent of the neocortex, independent of thinking. And if the feelings at the nexus of the tangled hierarchy of the two chakras—navel and heart—are truly autonomous, like thinking in the brain, what then? Our entire context of living undergoes a radical change.

From this new context, we can derive additional insight into how to attain emotional maturity. The navel chakra is where the body ego is located; our feelings of security, self-respect, and self-love reside there. The heart chakra is where love for the other is felt. And the bioelectric body associated with the heart chakra, together with the bioelectric body associated with the navel chakra, form a tangled hierarchy that can achieve a self-referential collapse of feelings in this navel-heart system—both self-love and other-love.

All this has to start at the level of feelings. When I allow the feeling energies between these two chakras to become causally circular—when the feeling relationships between the body individuality and the heart togetherness become a tangled hierarchy, an identity with the self of the navel-heart combination develops. Let's call this the tangled-hierarchical heart.

Feelings at the navel chakra are an expression of selfishness in terms of sexuality. Subjects I love become objects of sex when I identify with the body ego. Sex becomes a conquest. But unconditional love is the opposite of this. It is an expansion of consciousness. So when I realize that selfishness is giving way to expansion, it is a journey of discovering the self of the tangled-hierarchical heart. And when we integrate the self of the heart, which is based on feeling, with the self of the brain, which is based on meaning or rationality, we develop what can be called true emotional intelligence. Here, quantum physics helps us to understand the dynamics of how romantic love can become unconditional love, in which selfishness is transcended by the discovery of "the other." This can eventually lead to true emotional intelligence by integrating the self of the heart with the self of the brain.

Men often feel a stronger connection with their navel chakras; they likely have more self-love than other-love. On the other hand, women tend to be more connected with their heart chakras; their love is more likely of the other-love variety. This difference in connection is, I think, what is keeping humanity from experiencing the heart as a separate self and achieving true emotional intelligence. Our biological make-up contributes to this imbalance and the culture reinforces it. What we need is for men and women to defy biology, to defy cultural forces, and achieve a balance between these two. Men must develop other-love, and women must develop self-love.

In a previous chapter, we talked about the creative exploration of love. When men are sincere in their creative exploration of love, they discover the "otherness" of the one they are trying to love unconditionally. This is the beginning of other-love, which must then be harmonized with self-love. When women learn to explore love in the quantum way, they discover the otherness of the other, which leads to the discovery of themselves as individuals worthy of their own love. Their navel-chakra identity, self-respect, becomes strong as a result. Once this is achieved, establishing a tangled hierarchy is just a matter of harmonizing the two identities.

CHAPTER 14

The Meaning of Dreams

Our dreams are part of an ongoing life that we live in the meaning dimension. Dreams are investigations, explorations of meaning. In our waking lives, there is too much information occurring in the physical world that takes too much of our attention away from meaning. The ongoing storyline that we build for our lives because the physical world has this Newtonian habit of fixity also has the effect of distracting our attention from the meanings we are exploring, and the propensities we are developing for our explorations. The melodramatics that surround the storyline just take too much time and effort.

Dreams give us a hint of what meanings we are exploring, what propensities we are developing. Part of becoming enlightened through living in the quantum way—call it quantum living—is to pay attention to dream states. As we do that, and as we become sensitive to the transition between states of consciousness, we find more and more that we have what are called *lucid dreams*. In lucid dreams, we can use the dream state for developing solutions to problems.

During your dreams, you have experiences of different characters—people and things. These dream objects all stand for something—the meaning you ascribe to them in your waking life. For example, you may see your wife or girlfriend in a dream. But it is not your actual wife or girlfriend visiting you in her astral body. It is rather the meaning you ascribe to your wife or girlfriend. All these meanings appear in dreams in the guise of dream characters. In the dream, of course, you don't experience them at the meaning level. You experience them in the same way you experience them in your physical waking life, as if from habit. You experience a dream episode as if it were being enacted in physical reality. When you wake up, however, you can look at the dream from the point of view of meaning. That was not really your girlfriend, so what was it? What meaning do you give to that particular girlfriend? You give *this* meaning. She represents, symbolizes, *this* meaning to you. For instance, she may be a bit of a shrew. So she may represent the shrew part of you in your dream.

We untangle the meanings of every dream character when we do dream analysis. Another dream character may represent the miser in you, the part of you that refuses to be generous. Another may represent courage, your being brave. When we analyze the meaning of the dream symbols in this way, a pattern begins to emerge that can give insight into where you are in the quest for the meaning of your life.

Dream Lessons

In 1998, while doing research for a paper on quantum dreams through a grant from the Infinity Foundation and Institute of Noetic Science, I worked with psychologist Laurie Simpkinson. Although the paper we wrote together was never published, I included its main points in my book *God Is Not Dead* (2008). Dr. Simpkinson and I were able to show, through many case histories, that dreams are indeed ongoing reports of our meaning life. We also arrived at a new five-tier classification of dreams—physical-body, vital-body, mental-body, supramental, and spiritual dreams. Physical-body dreams consist of what is called day residue; they are dreams of events and people for which we do not find closure during the day. Vital-body dreams are those connected with suppressed emotional trauma. Mental-body dreams provide ongoing reports of our meaning life. Supramental dreams involve images of the collective unconscious. Spiritual dreams hint at our oneness with everything.

I will give some examples from my own dreams. When I was beginning to change from a materialist to someone who integrates the material and the spiritual, I had an ongoing series of purgation dreams that went on for months—dreams that involved lavatories, men's rooms, and toilets. The meaning conveyed, of course, was that cleaning toxicity from my system was very important to my process at the time. (For others, dreams of toilets can also mean letting go, not suppressing.) My system had a lot of materialist waste in it, and my dreams were drawing my attention to the need to clean it up—what Jungian psychologists call "shadow cleansing." At first, I didn't understand what was happening. When this meaning became clear, after a lot of dream analysis, I dedicated myself to cleaning up my system, and the purgation dreams stopped.

In the last dream in this series, two characters appeared—Ronald Reagan, a very conservative politician, and Jane Fonda, a very liberal actress. What stood out in the dream, however, was not who the characters were, but that, in the dream, they were dancing around detritus. Literally. The ground they walked on was full of excrement. I woke up with the feeling that labels of liberal or conservative were meaningless, and I was ready to give up letting someone else's opinion influence who I was. My slate was wiped clean, and I could get on with discovering my own opinion through real creativity—fundamental creativity. After this episode, I never had another excrement dream.

In the last chapter, I explored how to integrate male-female differences creatively. Carl Jung talked about this in a similar vein. In Jung's system, male potentialities of self-respect and self-worth appear to females as the archetype of *animus;* similarly, female potentialities of other-love appear to males as the archetype of *anima.* Jung advocated that males integrate the anima archetype in themselves, and that females cultivate the animus archetype.

As a man, at a certain period of my life, I was feeling emotionally dried up, very intellectual and very brain-centered. Then one night, I had a dream in which I was looking for water. I searched and searched and finally found a stream. As I got closer, however, I found that the stream was dried up. I was very disappointed. And then I heard a voice say: "Look behind." When I did, I was surprised to find that it was raining. I ran into the rain enjoying the way it fell all over my body. Then a young woman joined me—a delightful, pretty woman. We walked together for a while, thoroughly enjoying the rain and each other's company. When we arrived at what appeared to be her home, she said goodbye. Seeing the disappointment in my face, she added: "I am going to London for a while. But I will be back."

When I woke up, I immediately recognized the young woman as the archetype of my anima and felt excited about rediscovering emotional fluidity in my life. Of course, it did not come immediately—after all, she went to London. But she did come back into my life soon after.

Finally, I will tell you about my one unmistakably spiritual or "bliss body" dream. In this dream, I felt very joyful, and then I saw the source of my joy—a luminous man radiating joy that

I could not get enough of. And that was it; that was the dream. When I woke up, I consulted my dream teacher. He looked at me strangely and said: "Amit, don't you understand? You were dreaming of your own enlightened self."

Lucid Dreams

Lucid dreams are dreams in which we are aware that we are dreaming and are therefore able to guide the dream to some extent. My very first lucid dream occurred in the 1960s at a time when I was struggling with a particular equation connected with my work in nuclear physics. My memory of it is still quite vivid. The dream was about mathematical equations, so it was a bit technical. I was trying to find a way of applying some aspects of superconductivity to solve problems of the structure of atomic nuclei. But you don't have to understand about superconductivity to understand the meaning of the dream to me.

In the dream, I found myself thinking about equations, writing them down on what seemed to be a blackboard. Then I realized I was dreaming; something about the board was quite unusual. Whatever I was thinking, whatever change in the equation I was making in my mind appeared simultaneously on the board. It was a delightful way to work on equations, because I could actually see them without having to make notes. When I woke up, it took me only a few minutes to recapture the equation.

Let's return to the dream in which I heard a voice say: "The *Tibetan Book of the Dead* is correct! It is your job to prove it!" I don't know if this was a proper lucid dream, because I woke up as the dream was becoming lucid. But what I do know is that I probably would not have taken the subject of the soul and reincarnation seriously without this dream. In my book

The Self-Aware Universe (1993), I had the correct picture of the relationship between consciousness and matter, but I still did not understand the relationship between the mind and the brain. I was holding on to the illusion that mind is brain. But this dream inspired me to find the truth—that the mind is a different beast altogether, that the mind processes meaning while the brain just makes representations of mental meaning.

When we dream, the physical stimuli are brain noise. Just as the mind makes a meaningful picture out of a Rorschach image, so the meaning of all the symbols you see in a dream is the meaning you attribute to the brain noise. Therefore, in some real sense, all the characters in your dreams are *you*. The mental ego is therefore quite distributed and has little control in shaping the dream.

This loss of ego-control changes in a lucid dream, in which you are aware that you are dreaming within the dream. The dream ego is "boosted" by the waking ego in some sense, enabling you to guide the dream in certain intended directions. And you can use this vehicle of lucid dreaming to study the equipotency of your waking and dream lives. Who but Australian aborigines would believe that our dream life is as potent as our waking life today?

I think that lucid dreams have great potency for precipitating creativity, and that they can, therefore, be used for creative healing. Many lucid dreamers are interested in healing via the lucid-dream state, and several claim to have observed positive results from their efforts. In my book *The Quantum Doctor* (2004), I discuss how the mind can cause and cure health issues, as well as the importance of the "bliss body" and "creative sleep" in connection with mind-body healing. The bliss body is undivided consciousness—consciousness that is one with its possibilities, with no separation, no experience. It is beyond both waking and dreaming. In deep sleep, we are in the bliss body. Yet when we awaken, we remain the same. This shows that ego-conditioning is still controlling what possibilities we process while we are in blissful nonseparateness with our whole being. Creative sleep is sleep in which our ego-control gives way and quantum consciousness—you can call it God—can process new possibilities, possibilities of which creative experiences are made. This is quite different from lucid dreaming, but maybe we can call it "lucid sleep." When we experience this kind of sleep, we wake up highly creative, bubbling with creativity. Various traditions call these states of sleep by many exalted names, as we shall see in the next chapter.

Steven LaBerge performed experiments involving the lucid-dream state that indicated that the lucid-dreaming mind can affect (if only minimally) the physical sleeping body. There were also indications that lucid dreaming a particular task (according to activity monitored in the brain) was more like actually *doing* the task than just *imagining* doing it. And this may have important consequences for quantum healing.

Quantum healing consists of quantum leaping and healing the dis-eased structure of the mind. This, in turn, heals the vital energy blocks, which finally heals the physiology of organs. It is anybody's guess whether the creative shift of mental perspective actually percolates to the body physiology during a lucid dream. It is certainly possible, theoretically speaking. We need more data.

One dream I myself had stands out in my memory. In this particular dream, I knew that all the symbols in the dream represented me, because I was privileged to access the inside experience of my characters—not only what they were saying to "me" as a character in the dream, but also the "inside" of the images with which I explicitly identified in the dream. And

I was aware that I was dreaming, so it certainly was a lucid dream. It was very much like the mystical realization in waking awareness that we are all one.

Lucid Wakefulness

At a workshop I was presenting on the interpretation of dreams, a participant described the experience of waking like this: "Every morning, when I am suddenly awakened by my alarm, I get the same strange kind of feeling. I feel what it was like to be asleep and, in that one moment, I wake up."

I congratulated her and told her that she had learned to be very alert at that moment of waking, and was getting close to experiencing what I call "lucid wakefulness," in contrast to lucid dreaming. Lucid wakefulness can happen in the transition between sleep and wakefulness. These transition times are very good opportunities for having those creative "aha" experiences. Once you can focus your attention on these times, you are ready for quantum leaping.

At times in my own life, I have found these transitions very revealing. We can gain a great amount of insight in these moments of transition. In a way, when we wake up, we come back into our bodies. Because this often happens abruptly, most people don't develop the capacity to wake up with the kind of alertness that my workshop participant developed. When you develop a similar sensitivity, you will feel the discontinuity very clearly. If you do it even more sensitively, you can continue attending to the moment, knowing that it is a good thing that you are doing. Then these moments can be filled with tremendous insights about problems that you are trying to solve—creative objectives that you are trying to achieve.

CHAPTER 15

Enlightenment

Enlightenment is a subject about which those seeking personal growth have a lot of curiosity—and also a lot of confusion. Any time I do a workshop for psychologists and personal-growth seekers, I always begin by asking: "How many of you want to be enlightened?" Invariably, most, if not all, hands go up. When I ask why, I get answers like: "I want to be happy all the time," or "I want to be a person who can do no wrong." Everyone expects an enlightened being to be a totally transformed being, a perfected being.

In the spiritual literature, there are several types of experiences that qualify to some degree as enlightenment experiences. In Sanskrit, there are names for each of them. *Samadhi* describes the experience that the self is really the whole, with no individual self-nature in it. In samadhi, the self attains nothingness. No thing-ness. The Japanese calls this *satori*. *Moksha* refers to the final state of consciousness in which you have the choice to remove yourself entirely from the birth-death-rebirth cycle.

No Exit

Scientists—most of them materialists—generally do not believe that we can change, that we can ever overcome the drives of our negative emotional brain circuits. They also believe that we can never achieve perfection—we can never "get out of the game." Mystics, on the other hand, do believe in the attainment of perfection in this very human body. And if you attain perfection and have nothing more to seek—if you become enlightened—you are effectively out of the game.

But how is enlightenment part of the quantum worldview for the human potentiality? When I was a materialist, I didn't even believe in spirituality let alone spiritual enlightenment, even though I grew up in India where the culture and traditions are immersed in the concept. After my crystalization experience, I was writing a textbook of physics for beginners, trying to make physics interesting to students trapped in elementary physics courses. As I was writing about the three laws of thermodynamics, I adapted a Judy Collins song that was popular at the time.

The first law, the law of conservation of energy, states that the energy of the physical world always remains the same. You can't change the total amount of energy. You can't win. The second law states that, in every transaction, energy is degraded; its quality becomes worse. So, in truth, you can't break even. The third law states that it is impossible to attain a state in which thermal movement will give way to complete stillness. You can't leave the game.

As a materialist, I believed that these laws were, in fact, operative in the universe. But when I discovered that consciousness, not matter, is the ground of being, my views on such things changed radically. Why should I be bound by the laws of thermodynamics? Those are meant for purely material objects—inanimate objects—not for living beings.

In the seventies, before I discovered the quantum worldview, I studied a lot of Zen philosophy. In the Zen tradition, ten images—called the ten ox-herding pictures—describe the stages leading to enlightenment. In the first, you look for an ox lost in the mountains. Then you find it, bring it back home, and come down from the mountain. In the eighth image, there is nothingness, signifying the state of enlightenment. An enlightened state is achieved with the wisdom that reality is nothingness and the self is really no-self. But there are two more images after that. After enlightenment, the master goes down to the village and plays with the children. This seems to explain what happens to our being when we manifest or embody the wisdom attained in the eighth image.

I found it wonderful that the images did not end with the wisdom of nothingness, for what good is an experience, any experience, if behavior does not change? I have come to know a few "enlightened beings," and they have led me to believe that enlightenment is overvalued when the enlightenment experience itself becomes the final goal. Of course, I do see the experience of enlightenment as something very positive, but most of the enlightened beings I know don't seem to have come back from their exalted state as what I would call "transformed" beings. My confusion increased substantially when I myself had something that I thought could be called an enlightenment experience, because I knew I hadn't been transformed by merely having the experience.

In many spiritual traditions, however—especially in the branch of Zen called Rinzai Zen, which follows the Zen master Rinzai—the idea of enlightenment means having a sudden experience, much like a quantum leap, that is called satori. The confusing part, however, is that there is an expectation that, somehow, after such an experience, the person will also become an ideal human being—a transformed being who can love everyone and do no wrong.

This expectation does not usually bear out very well with what we see. We can follow people today, people who claim to have attained self-realization through satori or samadhi. We can follow them very precisely using GPS; hiding is impossible today. But if we did, I am sure we would be disappointed. The enlightenment experience by itself does not seem to produce transformation to perfection. Supposedly enlightened people even get caught in the usual scandals involving money, power, and sex. This has definitely affected people's belief in these ancient spiritual traditions.

So is the sudden experience that traditions talk about actually enlightenment? And if so, what is enlightenment good for if the expectation of perfection is not met once it is attained? What should we expect from our teachers who claim to be enlightened?

The Quantum Self

In the quantum theory of creativity, to investigate any archetype we have to go through the creative process. The archetype of self is no exception to this. We have talked about the creative process before, but here let's concentrate on the stage of creativity called the creative quantum leap. This is a discontinuous change in our understanding about an archetype. In the case of enlightenment, this is the archetype of self. In a creative quantum leap, something is discovered that we have previously never thought of or encountered. For example, in the experience of satori, this is the discovery that the experience is totally empty of all previous conceptions of the self. It is a discovery of emptiness, because all previous concepts of the self just fall away. Thus it really does represent a discontinuity in thought, in conceptualization. Before, you had so many concepts for the self; now, you have nothing. If the nature of the self is this nothingness, there is no object with which you can describe it.

And yet, there is now this "empty" self that you have realized. The quantum theory of creativity says that, in creativity, there is a further stage that you have to go through after the discovery of this real self, the "real self of emptiness." You entered the creative process with what we call an ego. Indeed we have talked about ego's storyline and persona; we have talked about ego's character. But then, after this quantum leap, we find that there is no such thing. The self is empty of all these concepts. And the quantum theory of creativity says that you have to manifest this no-self in your being. You have to manifest that emptiness of self in your life. This is why the Zen images show stages beyond satori—the stages of manifestation of this no-self self.

In the quantum worldview, there is a name for this no-self self. We call it the *quantum self*. Remember, in quantum dynamics, there is no fixity. So you can think of the quantum self as a dynamic self-experience that happens. But there is no place for this self to stand because there is no fixity. You cannot describe what it is, because you cannot stay in it long enough. We can call it by adjectives like nonlocal, or unconditioned, or creative, or tangled-hierarchical, or one—but that's the best we can do.

To identify with this quantum self is a proposition very different from identifying with the ego. Because we can only identify for certain with something that is fixed, how can we identify with something that is changing and comes only with momentary experiences? We can say that the self has to become like a fluid, changing shape depending on the container. Or we can say that it is an individual salt figurine melting away in the totality of the ocean. But these are all metaphors, not descriptions. What actually has to happen to anchor our being to the quantum self?

In the Hindu tradition, the enlightenment experience is called *samadhi*. This seems to be the highest experience possible, because you have experienced the nature of the self itself. But then Hindus ask the question: What is beyond experience? For example, in sleep, we don't have any experience. But sleep is not particularly transformative. After sleep, we wake up. We haven't changed. And it is easy to understand why that is. In normal sleep, we continue to process, albeit unconsciously, the same kind of repressed emotions and suppressed thoughts that we have a problem with in waking life. This is why most of our dreams are "day-residue" dreams. So when the ego identity falls away in manifestation, the attachment to the suppression-repression dynamic also begins to fall away.

But what does it mean to mature beyond the self-realization experience, beyond samadhi? What does it mean to manifest "no self"? The quality of what we call sleep becomes different, along with the quality of dreams. In this new kind of sleep—creative sleep—when we are one with the totality, we are no longer processing *only* the suppression-repression dynamic. In its place, we process wonderful new *possibilities* of consciousness at least some of the time. And when we wake up from this kind of "super" sleep or "creative" sleep, we wake up to profuse happiness instead of the suppression-repression dynamic—what we call "mood"—that often engulfs us upon waking from ordinary sleep. When we wake up from unconscious states of this quantum kind, our joy will certainly be reflected in our behavior in the community in which we live. More and more, we will leave behind mood and a perception of specialness, and we will live with the flow of life and ourselves become part of the flow. Hindu literature emphasizes the development of this capacity for such creative sleep, which it calls by another exalted name—*nirvikalpa samadhi*—samadhi without separateness.

There are two kinds of this nirvikalpa samadhi referred to in the experiential literature, just as there are two Zen stages that follow enlightenment. In the first kind, there is still a tendency to identify with the supramental, the archetypal values. In other words, there are still qualifications applied to the self-identity. This is the ninth Zen stage. In the tenth and final stage—the second kind of nirvikalpa samadhi—all possibilities are allowed. There is *no* distinction that this is a heavenly possibility and that is an earthly possibility. There is no preference. In the Zen images, this person is now an ordinary part of the community. There is no distinction left, no specialness. In other words, the Zen tradition centers on the behavior; the Hindu tradition centers on the state of consciousness—that creative sleep-like state in which enlightened beings spend some of their sleep time. The two traditions give divergent, complementary, descriptions.

Suffering and Commitment

Why are these expectations sometimes not met? Some people wade into self-realization investigations without proper preparation. In Buddhism, this basic preparation is expressed in Buddha's first law: Life is suffering. You are ready for the investigation of the nature of the emptiness of self only when you see, unequivocally, that life is suffering. You have to experience life as suffering literally. This means that even the creative exploration becomes suffering to you.

How can creative experience be suffering? When it becomes boring. When you have done this so many times before that you no longer have incentive to search creatively for the nature of the archetypes, other than the self archetype itself, the nature of the seeker. When only one question remains: Who am I that seeks? Only then, will you be ready and committed.

> While taking a morning walk, a pig and a chicken came across a restaurant serving breakfast. The sign outside advertised eggs and sausage. The chicken wanted to go in to eat, but the pig declined. The chicken insisted. The pig sighed and answered: "For you, it is just a contribution; for me, it is total commitment."

Clearly, the pig was not ready for that! And you can easily see from my behavior, especially if you watch me closely, that I am not ready for total commitment either. I am not ready to give up accomplishment and seeking. Clearly, the self is one archetype that I have yet to investigate seriously.

In Buddhism, you are not supposed to investigate the nature of the self unless you have reached the state of total boredom with life. Today, of course, we rationalize. How can life be total suffering? When endorphin molecules circulate in the brain, how can we deny pleasure and commit to suffering? We are on a pleasure "high." Although we are not prepared, we think that we are ready to investigate the nature of the self. But begun without proper preparation, the seeds of accomplishing this languish. Even though you seek satori, when you try to live a life of emptiness of self, you find that you still have things to accomplish.

For me, I want to be a teacher; I like to be the center of attention. I want to enjoy my enlightenment. But how can I enjoy my enlightenment if I don't let other people know that I am enlightened? All this simply sets up a dynamic in which the ego raises its head again. And whenever ego is present, there will be the risk of failure, because you are not going to meet people's expectation of a perfected ego-less human being. Again, we rationalize by saying: "Oh, enlightened people still remain a little non-enlightened. That's okay." No, it's not like that. It is just that people have not followed through to what is true enlightenment. They haven't reached the state depicted in the tenth Zen image. Quantum thinking removes all this ambiguity and makes things clear for all of us.

When people ask me if we can maintain the state of enlightenment over time, I tell them that this is not a valid question. It is not a question of maintaining the state of emptiness, because we can't. If we live in the waking state of consciousness, even part of the time, we will need to use the bathroom, for which we need the ego's expertise. Likewise, when we eat in public, we have to remember how to use a fork or a chopstick. If you want to live in society and be a teacher, even an enlightened master has to eat properly according to the expectations of the culture.

So living demands a certain amount of ego content. What we need are strategies to manage this content. The Hindu strategy is to spend more and more time in the unconscious—not sleeping in the usual sense, but in a special state of consciousness that I call creative sleep. Enlightened teachers in India spend a lot of time in this special state of consciousness—nirvikalpa samadhi. The Zen strategy is to live inobtrusively and become one of the many, without attracting attention and creating obstructions by unnecessary interactions. In this way, you avoid expectations because you are not interesting; you don't deserve attention. Quantum physics does not give us much to go on here either, other than the clue that, if you *have* to allow ego-conditioning into your life, keep it to a minimum. And by all means, only use the ego functions and try not to identify with the ego.

We have to learn to live in such a way that we don't need expertise that is demanded by others, only that which comes naturally, appropriately, if it comes. If I am your invited guest and am eating with you, you have an expectation of what I should or should not do. However, if I am just one of many—nothing or nobody special—then no one is watching me or how I eat. I can eat in whatever way comes naturally to me at that moment. I just go with the flow, as the spirit moves me. Then I am operating without calling on any ego conditioning. I become like a child who has no ego, who doesn't know how to behave. It is the attempt to conform that brings back the ego.

People sometimes ask me when I will seek self-realization, seek true enlightenment. Well, I am not yet bored of exploring the archetypes, or writing books, or teaching. But the self archetype is always out there, tantalizing me in the form of a recurring dream. In the dream, I begin at my old physics department and venture outside to explore other buildings—the arts, the humanities, psychology, religion. When I want to return to the physics department, I cannot find my way. Often the path ends in a discontinuity of sorts—an unfinished bridge, for instance. When I interpret this dream, I see the physics department as my "home"—my spiritual, true, quantum self-identity. To find it, I have to explore the self archetype; I have to make a discontinuous leap. The unfinished bridge is an invitation to make that leap.

In this chapter, we investigated how to bring about personal change. In the next, we discuss how to bring about complementary change in society. Buddhists recognize the importance of right livelihood. Scientific materialism, on the other hand, has narrowed down the professions so much that, today, it is very difficult to earn a living without getting distracted and to find satisfaction doing it. We have to change our social systems if we are to have any hope of finding the right livelihood for a person who wants transformation.

CHAPTER 16

Quantum Professor, Quantum Society

To expand our discussion to include a social perspective, let's use the example of a business person who is also a quantum activist and wants to do business according to quantum principles. Is this possible?

A business person is someone who has chosen to explore the archetype of abundance. In our current culture, however, it is very common for business people to adopt a very limited view of the archetypal journey—to become materially rich. Wealth at the material level is a valid representation of the abundance archetype. But it is not the only representation, and it is certainly not an inclusive one. In previous cultures, people remembered that an archetype is forever, so its representation must also be as permanent as possible. Material wealth in these cultures thus always had some sort of permanence. Real estate was viewed as wealth. Diamonds and gold represented wealth. In fact, the more permanent something was, the more valuable it was considered to be.

In our materialist culture, after we removed the gold standard and replaced it with unbacked paper currency, even the notion of wealth as something that was semi-permanent disappeared.

Not coincidentally, within a few years of abandoning the gold standard, we created mutual funds, and futures, and derivatives, and all sorts of value measures that are ephemeral and can plummet overnight. Derivatives are, in fact, abstractions of the idea of money that have become symbols of wealth, of abundance. So today, when people explore the archetype of abundance, the exploration has morphed into a process of simulating an archetype through a symbol just by the sheer brute force of consensus and information-sharing. If I claim to have information and the business community agrees that the information has value, then I have wealth. Some people have gotten very rich this way, on paper. Today, the peddlers of business information are the economists.

> Three professionals, a surgeon, a physicist, and an economist, are arguing about the world's oldest profession—the profession of God. The surgeon says, "God must be a highly skilled surgeon to have made Eve out of Adam's rib."
>
> The physicist laughs and says, "Ha! Everybody knows God is a chaos theorist who brought order out of all that initial chaos."
>
> The economist smiles and slyly says, "Yes. But who created the chaos?"

Quantum Business

When quantum activists engage in business, they value the deeper meaning of an archetype like abundance. They recognize that, when an archetype is investigated and explored even in a general way, the exploration automatically leads to other relevant archetypes like love or goodness, which are to be embraced and cultivated. They not only develop a respect for ecology, but also for deep ecology—ecology of the psyche. Psychological awareness develops concurrently with ecological awareness. With renewed interest in this "deep" ecology and psychological investigation, creativity returns and naturally leads to a respect for meaning. Quantum business people accept supramental values along with creativity's return as a fully comprehensive force, not exclusively as information-processing. And they include these values in an ever-increasing technology—material and nonmaterial.

But quantum activists are not always welcome in the business world. People around them may not be able to exist in harmony with them. On the other hand, the people around them may be influenced by their way of thinking. To succeed in the business world today, quantum activists must have ideas that are innovative and that always satisfy the bottom line. Today, there are enormous possibilities for starting new forms of business and developing nonmaterial technologies that will be more conducive to the quantum worldview.

Although the idea is still very new, society is opening to the possibility of nonmaterial technology—for instance, in healthcare. Today, allopathic medicine dominates our healthcare system. Thanks to accumulating empirical data, however, alternative medicine has now made some inroads into the field, in turn creating business interest. And alternative medicine, much of which uses nonmaterial forces like vital energy, is clearly aligned with the quantum worldview.

Quantum Medicine

With the help of the quantum worldview, we now have integrative medicine, which integrates both alternative and allopathic medicine. This has opened up many possibilities for healthcare professionals and business people to collaborate in a field called integrative health management. Currently, healthcare practitioners in both conventional and alternative practices are, for the most part, empirical practitioners who are very limited in theoretical knowledge. Allopathic medicine, of course, has no theory and hardly any philosophy of disease other than the tired old germ theory. Whether a particular medicine or a particular surgical procedure will be effective is viewed as entirely empirical. Clinical studies decide the efficacy of any treatment.

On the other hand, alternative medicine relies on ancient theories. But scientific materialism has lessened our faith in these theories, even among the practitioners themselves. More than three quarters of our medical dollars are being spent on care of the elderly treating chronic diseases like cancer and heart disease. But we know that, except for emergency situations that require surgery, chronic disease is best treated by alternative medicine. Allopathic treatment is not only costly, but also fundamentally ineffective. It works to alleviate symptoms in the short term, but produces side-effects in the long term that bring about further deterioration in health. If the disease does not kill the patient, the side-effects of allopathic medicine often make life so miserable that it is not worth living. In this situation, integrative quantum health management can not only co-join theory and experiment, but also open up the possibility of preventive medicine.

Preventive medicine will eventually enlarge the scope of vital-energy practices, opening up additional areas of activity like vital-energy-sensitive ecology, the efficacy of maintaining positive health, and creating and continuing positive mental health. The information overload that is dragging down the human condition is like a form of pollution. While we need a modicum of information to explore meaning, constant bombardment with information can bring about conditions, like attention deficit hyperactive disorder, from which many of the millennial generation are already suffering. New areas of business activity and new approaches to healthcare will revolutionize the whole economic scene and substantially improve our quality of life if we allow new domains of technological exploration like vital energy, mental meaning, and spiritual values to really take hold.

Quantum Banking

New financial models have surfaced in some places. For example, the Grammin Bank of Bangladesh merged a traditional model with a new model to come up with a sound business practice that improved the lives of many—micro-investing. But, while this model worked on a small scale, it is struggling to succeed on a larger scale in countries like India. The original idea was modeled for a small-scale economy—typically for a village, to cmpower women. It has been very effective in these venues because, indeed, village women in these cultures have the suitable value structure to achieve business success. All they needed was to be empowered.

Once other people got the idea that they could make this particularized structure work on a much larger scale and turn a profit for themselves, the idea faltered. Once large-scale entrepreneurs entered the picture hoping to capitalize as intermediaries without understanding the value structure on which the idea was built, the model was doomed to fail on a business, as well as on a humanitarian, level. These intermediaries simply did not have the same value structure as those who were directly using the investments of the banks—the village women.

In order to make this model work on a larger scale, we have to change economics itself. We have to change the value structure of the society itself. We can do this by introducing values like love and happiness as economic commodities and by establishing the abundance archetype as a permanent fixture in our business dealings.

Human Capital

Let me tell you a story. My wife and I were walking by the river in a small town called Rishikesh in India. Suddenly a door opened and a man came out. "Laughing Baba is giving a discourse," he said. "Would you like to come and listen?" The Indian word "baba" means father, and spiritual teachers are usually referred to by this title. But the adjective "laughing" intrigued us, so we went in and sat down. Baba was teaching the *Bhagavad Gita*, a very famous Hindu treatise, speaking in the Hindi language, which I don't speak very well. So my attention wandered. I started to get bored and began looking around. As I did, I noticed something rather strange. Everyone in the audience had a little smile on his or her face. And then I noticed that I, too, was smiling. And suddenly I knew why he was called Laughing Baba. In his presence, happiness wells up and makes everyone smile. If we could cultivate people endowed with this kind of presence, we could sell happiness for a profit!

Today, our workplaces are places of dissatisfaction and unhappiness. But what if we were to change that? Beautification programs construct beautiful surroundings so people can enjoy aesthetic environments. What if we were to introduce these intrinsically joyous people into our workplaces to build happiness into our places of business. Moreover, this could generate revenue, because the business would doubtless experience increased profit through increased productivity. It is a fact that increased satisfaction or happiness leads to increased productivity.

What should we call these Laughing Babas, these happy people who give off happiness like a flower gives off fragrance? Human capital, of course! In view of the fact that machines are taking over many routine jobs, isn't it time we consider educating people to become human capital instead? We are, of course, in the very beginning stage of this kind of development. My book *Quantum Economics* (2015) proposes just this sort of thing, and I am just beginning to reach out to businesses with these sorts of ideas.

Vital Marketing

Scientific materialism, despite its many shortcomings, produced some very powerful ideas in business economics. One of these is the use of marketing to create and shape interest in a particular consumer product. If the product is really unnecessary, then marketing becomes a dishonest process. But if a particular product satisfies in any way, marketing techniques can help to bring a particular need to the forefront of people's consciousness.

Today, people really have very strong needs for vitality, but they are not familiar with the assessment of feelings—feelings in the body. Because of this, they generally cannot comprehend that we can learn to experience feelings in the body. Through simple exercises, we can even learn to augment these feelings. At the same time, people don't realize that we already have the technology to restore vital energy in some of the things we consume—products that lost that energy because of the way they were manufactured.

Take the case of perfume. Over the ages, people have used natural products like roses for fragrance. Newlyweds discovered that roses strewn on the bed enhanced their amorous and romantic feelings. This effect relied not so much on the material scent as on the vital essence of the roses. When modern chemistry entered the picture, however, chemists began extracting the material essence, the molecules, and making perfumes that kept the fragrance of the flower, but lost its essence. Through marketing, these perfumes are now sold at very high prices. But women eventually discover that wearing them doesn't really do much to invite romance. They smell good, but that's it.

Today, however, we have the technology to put the vital energy of roses back into the perfume. Imagine the marketability of a perfume that actually delivers results—that actually enhances romantic interest. And its not just women who will appreciate this. Men will want to wear this perfume as well. In order to make this product and its marketing effective, however, we have to train people to feel energy in their bodies. This will require a two-prong effort. The first will provide general education for people to become more aware of their feelings as vital energy in the body. The second will bring products into the market that contain vital energy and prompt consumers to buy them.

Materialist science depends on technologies like computers and cell phones. By contrast, quantum science is introducing new technologies that are very powerful and very rewarding. Material technologies can start out making you feel great. With time, however, the experience becomes jaded and eventually boring, until finally, you only get a mild form of suffering from them. And you keep on wanting "new and improved" material products to keep your pleasure buds going. The technologies of consciousness do just the opposite. Initially, they feel like medicine, hard to get down. But with time, you begin to look forward to the experiences you get from them. You get more and more satisfaction. And you don't tire of them.

Of course, in order to sell products like this *en masse*, we also need a third thing—quantification. We need to be able to measure and quantify vital energy and its effects. We are currently developing techniques for this, and vital-energy technology is proving to be a very productive application of the quantum worldview.

As Abraham Maslow said as early as the 1960s, when our survival needs are satisfied, we go for higher needs. In an affluent society, especially as the worldview changes and information-processing gives way to feelings and meanings once more, there will be a market for this vitalized rose perfume. Women will flock to purchase it, just as today they flock to buy the sixth-generation cell phone. So this is a case where quantum activists will work very well with business people. In fact, I am actively looking for entrepreneurs who will start producing these revitalized perfumes.

The Quantum Economy

How should we think about money in the quantum worldview? When we used to hunt animals to feed ourselves, we had no need for money. When we became an agricultural society, we started to trade or barter. Then we developed coins and formed money-based economies because they worked better for long-distance trading. But this sort of economy is not doing very well now, is it?

In truth, the current world economy is running very poorly. Our economic model fundamentally depends on ongoing expansion. But how can indefinite expansion take place in the finite ecosystem of the material world? Sooner or later, the need for sustainability will defeat us. If we don't prepare ourselves and face this possibility, we will be doomed. The shortage of resources will catch us unaware.

Fortunately, the quantum worldview, when applied to economics, can help. In a quantum world, all experiences are equal—all experiences are important and worth having. Everything is needed. All experiences fulfill a human need, a deep human need. Fundamentally, economics is the matching of these needs to the gifts some people have to make products to satisfy these needs with the available resources. Adam Smith's great idea—that gifted people produce and needy people consume, both with their own self-interest in mind—suggests that, if you allow the invisible hands of the market to operate freely, without constraint, then needs and resources, production and consumption, will find an equilibrium. Demand and supply will match. And resources will be allocated properly.

This idea worked for a long time. Its fundamental truth is grounded in the notion that people have to be free to pursue their own selfishness—their selfish needs as well as their selfish gifts. But we cannot forget that human beings are also social beings, which introduces the idea of collective need and the concept of social good into the equation. Adam Smith's hope was that, if capitalism were allowed to develop freely, it would result, not only in individual good, but also in social good. Generally speaking, however, we find that only the individual part of this equation holds true.

The idea that capitalism will produce social good in addition to individual good indeed has not proven true in most capitalist economies. The economics of John Maynard Keynes introduced the *ad hoc* idea that government has to impose social good through governmental intervention in recession times.

All advanced economies today—Japan, the United States, and Europe—have introduced this kind of social safety net. But even with this, capitalism is not working. The safety net fails because its cost seems to spiral out of control, making it difficult to support by any government. Add to this the perpetual boom-and-bust business cycle that is a fundamental feature of capitalism, and it is easy to see why capitalist economics as it is now practiced does not work.

So why not try introducing a quantum worldview into the mix? When we introduce elements of a subtle economy—products like vital energy and mental meaning, or even spiritual values like happiness—the boom-and-bust cycle disappears. Forget that these subtle products cannot be produced in bulk right now, and know that, in time, with a little effort and a lot of creativity, they can.

If we bring mental meaning, vital energy, and spiritual values—even happiness—into our economies as commodities to buy and sell, we can make a major dent in the problem of sustainability, because these dimensions of our experience are infinite. Their market can be forever expanded. There is no limit to how much love you can produce or how much love you can consume. And, in fact, when you consume quantities of love, you don't need to deal with the absence of love—you don't need to go on a shopping spree and buy material goods to fill up the emptiness of your life. You can reduce your material needs. And sustainability can be achieved.

Of course, different people have different needs, and different groups of people also have different characteristics and different needs. Countries like Japan and America have different cultures, and companies like Sony and Panasonic also have different characteristics. While corporations tend to form their own internal cultures, since the corporation is itself comprised of individuals who have grown up in the larger culture, some outside cultural influences cannot be avoided. But as corporations become more and more multinational, these differences are becoming secondary. Big, multinational corporations basically belong to the same corporate culture. It doesn't matter whether Pepsi Cola is run by someone from India or someone from France; it doesn't matter whether the CEO is a man or a woman. It makes no difference, because managers in multinational companies have to surrender their individual cultural and gender biases.

This is the result, I think, of scientific materialism, which homogenizes people and cultures. Materialism itself releases the power of the brain circuits, so to speak, because it engenders the belief that everything we are or can be is what we are already. The sum of our character is already all in our brains; it is not something that we can develop by virtue of intuitions and creativity. Scientific materialism has no room for intuition and creativity. You are what you are. You cannot change. All you can do is continue making permutations and combinations of what you already are.

If multinational corporations remain a permanent fixture on the business scene—and strong forces toward globalization certainly suggest that they will—then our challenge as quantum activists is to bring these corporations into alignment with the quantum worldview. This is certainly not hopeless. After all, Mitt Romney once famously told us that corporations are people. Well, if this is true, corporations need to humanize themselves. And that means they must adapt to the quantum worldview.

The present state of corporate culture is well-depicted in a *Dilbert* cartoon. Dilbert tells his manager that he's concerned that his personal goals don't match the corporate strategy. He says, "I would like to be happy." He asks his manager what the company wants. The manager replies, "Nothing along those lines."

Isn't that the truth with most companies today?

Business researcher Richard Barrett, in his 1998 book *Liberating the Corporate Soul,* found a lot of progress in how certain corporations are trying to align their corporate values with their employees' personal value structures. Moreover, he found that this was improving productivity and the bottom line.

Certain aspects of the new quantum economics also suggest that the future will embrace more of the kind of economy that Gandhi and Schumacher propounded—small is beautiful. Some aspects of economics in a quantum worldview suggest that perhaps there will be a movement toward an economy in which smaller units will function better, because they will become more human-oriented. And, in fact, quantum economics proposes the development of human capital as its major focus. People themselves have investment value because they can develop their own gifts and talents to satisfy other people's need for vital energy, mental meaning, and supramental values. This is an idea whose time has already come, as the robotization of technology removes more and more jobs from the human labor market.

Quantum Politics

I began this book with comments about political polarization. Obviously, the integration of worldviews enabled by the quantum worldview will remove polarization from politics as well. But that is not the only problem we face in our politics today.

Today's politicians are elitists. They sabotage democracy instead of propagating it and making it more accessible. In a democracy, a politician's job is to use his or her influence to empower people so that more and more citizens can participate in the political process. But today, politicians are more interested in using their power to dominate people and perpetuate elitism. Democracy is rapidly giving way to oligarchy—government by a small ruling class.

What guidelines does the quantum worldview give that political leaders can follow? The liberals have to take the lead on this. This whole problem of shrinking democracy due to rampant elitism has arisen because liberals, by and large, have bought into a worldview defined by scientific materialism. Any thinking person who is a nonscientist should find it quite appalling that human behavior, political behavior included, can be understood through material laws. However, almost all of today's liberal political leaders are trained at institutions of higher education that have sold out to scientific materialism. The arts and humanities, not just spirituality, have order and regularity, but these regularities are not lawful like scientific laws. Instead, they play out storylines that we depict with what we call *mythos*—mythology.

One of these mythical storylines is called the hero's journey. In the first stage of the journey, the hero sets out to find truth or wisdom, including political wisdom about what to do with power in a democracy. In the second stage, the hero, through much trial and tribulation, discovers wisdom. In the third and final stage, the hero returns triumphant, ready to use his wisdom to empower people.

In another mythological story, the myth of the Holy Grail, there is something wrong in the kingdom. At first, our hero sees this, but says nothing because of his socio-cultural conditioning. Only after much work (the hero's journey), does the hero acquire enough courage to ask: What's wrong here? And the kingdom is healed.

Of course, there are political leaders who constantly talk about the error of elitism and call for a return of political power and economic well-being to people. But they never exactly tell us how to address these issues except along party lines. In America, Democrats want big government to solve the problem of elitism; Republicans want to do it by cutting taxes—ironically, mostly for the rich elite. Both remedies end up promoting elitism. Obviously, today's politicians are very good with promoting fake solutions. But the truth is that they don't really see anything wrong with elitism, since they themselves are members of the elite.

The challenge is to bring those mythological storylines back into our political world in their true essence—not as fake promises, but as fact. But our leaders are confused because of their incomplete worldview, so promising is the best they can do. As a result, we go through the ritual of leaders promising to bring "real change" every four years, but, of course, nothing much changes. Only the quantum worldview can reintroduce the myth of the hero's journey into our politics.

Quantum Education

To implement that view, we will have to bring real change to the higher education establishment. But won't that take at least a couple of generations, you ask? Not with quantum principles in play. In the quantum worldview, the world is not run entirely by laws. It can't be. In quantum science, within the primacy of consciousness, matter is just hardware. We use matter to make the software of consciousness in the form of the self, but also of our subtle experiences in the form of brain memory and body organs and their modifications. The material hardware indeed follows physical laws. But, just as with computers, the hardware can tell us nothing about the behavior of the software.

In consciousness, we use the software of our subtle experiences to map our mental storylines and process them consciously. These storylines have order. If they did not, there would be no arts or humanities to learn about. The order comes from guidelines given by the supramental archetypes. Our mythology is the history of the play of these guidelines.

Will future political leaders ever get on the same page with the quantum worldview? I think they will. The humanists—and who can deny that the humanistic influence on liberalism still continues—agree that mythology is as important as law in shaping human drama, and that includes political drama. As William Irwin Thomson wrote not so long ago: "Mythology is the history of the soul [our supramental body]." Thus, politics has to bring back humanism and mythology. And the quantum worldview and quantum economics can make this possible.

The quantum worldview is going to change education radically. Today, our education system has ceased to provide what used to be called a liberal education. In fact, the whole meaning of the word "liberal" has been corrupted under scientific materialism. It now means "supporter of science based on scientific materialism." Originally, liberal education meant an education that liberated you from dogma. And that meant supporting science against religious dogma. But conventional science has now embraced scientific materialism, which is also a dogma—materialism that excludes spirituality, as opposed to spirituality that excludes the material.

We need integration to balance the material and the spiritual, and the quantum worldview gives us that integrative worldview. And we have to bring this worldview to education. We have to liberate secular education from scientific materialism, just as we did 400 years ago when we liberated education from religion. And, of course, the task of liberating spiritual education from religion still remains as well. Religions, even the Eastern religions, have a prejudice against the material world. But the quantum worldview requires both spirituality and materiality, and we must work to bring this integrative spirit into not just our educational system, but into our society as a whole.

Most people still consider spiritual things to be somehow unreliable—"way out," or "on the fringe." This has always been a barrier to spiritualizing society. But what the quantum worldview tells us is that material things are also "way out." When you question the reliability of consciousness and its role in the world, you must realize that all material realities are just possibilities. It is its interaction with consciousness that gives matter the concreteness that we experience as a sensation. When we experience the mind, we experience meaning. When we experience matter, we experience only its material qualities. We need this gross material experience, because we need the concrete to differentiate the subtle. That doesn't make the subtle superior, nor does it make matter superior. We need both—mind and matter.

Take a very satisfactory gross material product like marijuana. You inhale marijuana. It goes into your brain cells. This is a chemical interaction. But unless there is a subject experiencing the result of this brain-pot interaction, there is no value in it. The feelings generated by the opiate receptors being filled by the marijuana molecules comprise the experience that gives value to pot. The relaxation you get depends entirely on whether or not you are there to be relaxed. A robot will never need marijuana.

There is some evidence that even the millennial generation is tiring of information-processing and is coming back to human meaning and values. We can hope that this quantum way of living is going to be integrated into the whole society eventually. I think that's the power of the quantum worldview.

At a recent conference on energy healing at a spiritual center at Pyramid Valley in Bangalore, I gave a talk along these lines that was well received. Afterward, I was invited for a private session with the spiritual Master of the place, Brahmarshi Patri. Frankly, I was venting my frustration with the current politics of competing worldviews, pointing out how blinded the Western public was by its culture. I pointed out that 40 percent still see science as synonymous with scientific materialism because, historically, that philosophy has brought the West much power over the rest of the world. Another 40 percent holds on to the idea of God as a super being—a king of kings, an individual power, not a cosmic force. "Caught between these two restrictive worldviews," I complained, "the Western mind refuses to embrace the quantum worldview because, to it, it sounds too Eastern to trust."

The Brahmarshi listened intently. Then he surprised me with this suggestion: "Why don't you start a university that will teach everything—science including health, social science including business, arts and humanities, and spirituality based on the quantum worldview, integrating science and values. It could become a university of spiritual transformation."

Acting on his insight, I and about thirty others—Americans, Europeans, and spiritual-minded Indians—are busy doing just that. We are developing a post-graduate university of spiritual tranformation that grants master's and doctoral certifications. Our venture is called the Quantum Life University and our plan is to open our doors as early as July 2017.

Quantum Activism: Some Final Words

The central idea of quantum activism is not that we must advocate a dogma, but rather that we must learn how to be *free* of dogma. Quantum activists must learn to be creative individuals—literally the originals that we previously called geniuses. The quantum worldview teaches us that anyone has the potentiality to be a genius, so the movement does not depend on any particular individual. The system itself produces many such "individuated" people—creative individuals.

Some will achieve this individuation—this originalization (there! I just made up a new word)—quickly. But what about those who have no interest in quantum activism, or in any kind of activism for that matter? What about people who are preoccupied with bare survival? Remember the fable of the hundredth monkey? Perhaps that is the way that quantum change will arrive. When a threshold is reached, the whole society will change and become a quantum society.

The idea of a threshold of change is based on the Lamarckian theory of evolution—the inheritance of acquired characteristics. Evolutionary data regarding instincts supports this idea, says Sheldrake. I have relied on this idea as I theorized about how animal instincts of pure feelings become negative emotions in our brains. Or how the Jungian collective unconscious was created. But the theory crucially depends on quantum nonlocality, on people living in a society that has activated nonlocal connection among its members. In these societies, nonlocality is not just a potentiality. This was the case in societies of vital mind in the era of garden agriculture, and this is the case in tribal societies even today. But with the advent of the rational mind, we have become disassociated from one another; we have no nonlocal connection. I have suggested the Internet as a remedy, but let's face it. It's quite far-fetched to think that we can create a tribal society through the rational mind.

Is there an alternative? I think quantum economics gives us another way to encourage individual changes to permeate the whole society gradually. Quantum economics is based on the production and consumption of subtle energies. If you consume positive vital energies, like the energy of love, you soon become interested in becoming a producer yourself. And you begin the process of transformation. Of course, this sounds like—and probably is—a very slow process. In view of our current critical problems, don't we need to hurry?

Well, yes. But the idea of the threshold is very old. It existed in Buddhism long before it was developed by Lamarck or Sheldrake. Unfortunately, it is an elitist idea. It relies on leaders and followers. And that is not the way the quantum world works. The quantum worldview is against all forms of elitism. So however much the promise of the hundredth monkey attracts us, it may go against the grain of the movement of consciousness, which, I think, works strictly according to the quantum worldview.

Indra's Net

Consider the concept of Indra's net. Indra is the ruling god in heaven according to Hindu mythology, and Indra's net is an interconnected web of relationships. We are recapturing this kind of concept to describe reality today. Quantum physics has given us the concept of the domain of potentiality, from which the domain of actuality—of appearance, of manifestation—comes. And this domain of potentiality is an interconnected consciousness, like Indra's net. How does it manifest in our experience?

This is what the book is about. How can we make ourselves and our society more and more profound in terms of the experiences that absorb us, and give our lives meaning and satisfaction? Many of us get side-tracked. Many say that we have too much suffering. Many see problems. But what we must see is that there is always potentiality to solve the problem.

How can we get to potentiality? How do we become creative? How do we access the nonlocal domain of potentiality that we call consciousness? How do we evolve to make this available for the whole of humanity? How do we change our businesses? How do we change our economics? How do we change the power structure and elitism in politics? How do we change our health and healing systems? How do we change our education system? How do we change ourselves most of all? I hope, dear reader, that considering these questions and the answers that the quantum worldview gives us has helped you to love, to engage meaning and values, to research, to investigate, and to explore.

Glossary

aikido: Japanese martial discipline that uses gestures and arm-and-hand movements to raise energy from the root chakra all the way to the crown chakra.

allopathy: "mechanical" model of medicine that grew out of scientific materialism, based on assumptions that all disease is a function of physical causes and effects.

anthropic principle: principle that states that the world is designed to move toward establishing manifest embodied consciousness.

archetype: object of the inner experience that provides the context for intuitive and creative thoughts and feelings, our most elevated thoughts and most noble feelings. Platonic in origin; Jungian archetypes denote representations of Platonic archetypes that have become part of our collective unconscious.

arrow of time: appearance of directionality of time from the past to the future.

aura: biophysical electric body, different from the biochemical body of which the organs are part.

autopoiesis: the self-making of life posited by Humberto Maturana; in a living cell, the tangled hierarchy of DNA and protein generate the self-making loop that supports life.

Ayurveda: system of medicine with historical roots in the Indian subcontinent; a type of complementary or alternative medicine based on the the healing of the vital body and its connection to the physical.

bliss body: undivided consciousness; consciousness that is one with its possibilities, with no separation, no experience. In deep sleep, we are in the bliss body.

chakras: seven vital energy centers each located near one or more major organs and associated with those organs' biological functioning and feelings experienced through the vital energy of movement of morphogenetic fields associated with those organs.

Chi Gong: Chinese martial discipline that uses gestures and arm-and-hand movements to move vital energy; a variation of tai chi.

circular causality: tangled hierarchy that occurs in a self-representing system.

creative sleep: sleep in which our ego-control gives way and we can process new possibilities of which creative experiences are made.

co-arising, dependent co-arising: Buddhist concept of the co-arising of subject and object from the unmanifest into manifestation, in which the subject does not create the object, nor does the object create the subject; they are co-created. Similar to the quantum collapse of unconscious potentiality into subject-object awareness.

collapse: transformation of a wave of possibility into a particle of actuality.

collective consciousness: set of shared beliefs, ideas, and moral attitudes that operates as a unifying force within society.

complementarity principle: principle developed as part of the Copenhagen Interpretation that resolves the wave-particle paradox by claiming that quantum objects are both waves and particles whose aspects can only be revealed in discrete measuring experiments; both aspects never show up in the same experiment and are thus complementary.

creative quantum leap: discontinuous change in our understanding of meaning.

dharma: particular learning agenda in a life, its incarnational purpose.

delayed choice, delayed collapse: event in which you remember the entire chain of events that consciousness chose and collapsed prerequisite for the present event, going backward in time all the way to the potentiality that started the causal chain.

discontinuity: examples are creative experiences that come as a surprise; "aha" moments.

DNA: one of the two essential molecules in a living cell; the other is protein. DNA is needed to make protein; protein is needed to make DNA.

domain of potentiality: domain of quantum waves of possibility, quantum objects in their original form, as differentiated from space and time, where communication is signal-less, nonlocal, and instantaneous. Not to be confused with the domain of potential, which has other meanings.

downward causation: causation by conscious choice from potentiality into actuality, so named because it emanates from a higher consciousness transcending the ego; causation behind collapse and the power behind the event of collapse.

dualism: the idea that anything nonmaterial must exist as a separate object.

ego, ego-consciousness: consciousness that is the result of conditioning.

feelings: movement of the vital body; the energy we experience through feelings is vital energy.

fixity: state of minimal quantum movement that allows us to make representations of our subtle experiences using matter.

fundamental creativity: type of quantum leap in which we follow through on our intuitions all the way to the essence of an archetype, make our own mental representation of our insight into the archetype, and eventually develop it into a product that others can appreciate.

God: in the quantum worldview, the causal agent of quantum consciousness.

involution: limitations imposed on the potentialities of consciousness.

karma: belief that the sum of actions in this life determines future existences; from the perspective of quantum science, karma is nonlocal memory conditioning from past lives.

kundalini: Sanskrit word meaning "coiled up energy"; the new movement of vital energy or life force activated in a kundalini awakening.

kundalini awakening: personal awakening in which prana (vital energy) moves from the lowest chakra to the highest; in quantum terms, new feelings at these chakras remain in potentiality until a sudden quantum leap of awakening takes place.

law of absolute zero: third law of thermodynamics that states that the entropy of a system approaches a minimum as the temperature approaches absolute zero; in other words, it is impossible to attain a state in which thermal movement will give way to complete stillness.

law of conservation of energy: first law of thermodynamics that states that the energy of the physical world always remains the same.

law of entropy: second law of thermodynamics that states that, in every transaction, energy is degraded as order gives way to disorder or entropy.

lucid dreams: dreams in which we are aware that we are dreaming and are therefore able to guide the dream to some extent.

lucid wakefulness: state of awareness that can occur in the transition between sleep and wakefulness.

modernism: dualistic worldview in which mind and matter are separate.

moksha: final state of consciousness in which you have the choice to remove yourself entirely from the birth-death-rebirth cycle.

monad: term from Theosophy meaning the entity that survives the physical death of a human being.

morphogenetic fields: blueprints for biological form and function-making for the organs of the physical body that play out the vital functions in space and time.

nirvikalpa samadhi: samadhi without separateness.

nonlocality: signal-less communication that occurs via mediation through consciousness, the domain of potentiality.

object: material construct of our outer experience; by contrast, archetypes are objects of the inner experiences we call intuition.

observer effect: principle that a possibility wave of any given object or event changes into actuality only when an observer experiences it.

particle: one-faceted object actualized from a wave of possibility.

pranayama: practice that uses breath control to carry air all the way to the brow chakra and then down all the way to the navel chakra using deep inhalation.

purusha: Hindu concept of the realm of the potentiality of subject-consciousness.

prokriti: Hindu concept of the realm of the potentiality of objects.

protein: one of two molecules essential to the living cell; the other is DNA. DNA is needed to make protein; protein is needed to make DNA.

psychosomatic illness: errors in meaning-processing that can result in serious physical disease.

quantum: irreducible discrete quantity first used with this connotation by physicist Max Planck to denote the idea that energy exchange between bodies can take place only in terms of discrete units.

quantum consciousness: the ground of all being and the source of downward causation; consciousness that splits itself into a subject (that experiences) and an object (that is experienced); a nonlocal objective cooperative process.

quantum economics: use of the subtle, including vital energy, meaning, spiritual values, and happiness, as commodities in economy and business.

quantum leap: discontinuous transition; when an electron jumps from one atomic orbit to another without passing through the intervening space, it is a quantum leap. In a creative experience, when we jump from the known to the unknown without going through intermediate steps of thought.

quantum love: when we use quantum principles to explore the archetype of love.

quantum measurement: measurement that converts a quantum possibility wave into a particle; measurement that requires both the perception apparatus and the memory apparatus of the brain.

quantum measurement paradox: circular causal loop generated by the idea that the existence of the observer's brain requires collapse, while collapse requires the observer's brain.

quantum memory: memory that resides outside of space and time, not in the brain; nonlocal memory.

quantum monad: ongoing dynamic locus of surviving nonlocal memory for an ongoing chain of individual incarnations; ongoing dynamic locus of quantum memory that survives the death of individual incarnations.

quantum self: self or subject associated with an immediate experience, like intuitive experience; unconditioned self, as opposed to the conditioned ego self.

reductionism: interpretation of the way the world is structured in which the micro makes up the macro, and the macro is thus reducible to the micro.

shunyata: state of consciousness that transcends both subject and object in which potentiality is recognized as "no thingness" (nothing-ness).

satori: experience that the self is really the whole, with no individual self-nature in it; Hindus call this *samadhi*.

samadhi: experience that the self is really the whole, with no individual self-nature in it; the Japanese call this *satori*.

situational creativity: mental process in which we creatively interpret archetypes from the experiences of others, then receive a secondary insight into the archetype in the context given by that vicarious experience.

Siva (Shiva): one of the three major deities of Hinduism, called "The Auspicious."

space and time: opposite of the domain of potentiality, in which communication requires local signals that move through a locality; domain of manifest objects; the domain of particles.

subtle body: the conglomerate of nonphysical bodies pertaining to our internal experiences, namely the vital body, the mental body, and the supramental or archetypal body.

subtle realm: what we experience internally, as opposed to matter, which we experience externally. Matter is gross, fixed, and semipermanent; the subtle realm is forever changing.

supramental: another name for the world of archetypes, suggesting that the archetypes provide context for mental meaning and are therefore beyond the mind.

synchronicity: when two events—one in the physical world and the other in the mental world—are correlated through the meaning that arises in the mind.

tai chi: a Chinese martial discipline that uses gestures and arm-and-hand movements to move vital energy.

tangled hierarchy: a hierarchy of circular causality between levels in which the two levels are causally tangled. A causes B, and B causes A, *ad infinitum*.

transmodernism: modernism based in the integration of the quantum worldview and modernism.

upward causation: Newtonian worldview concept of a material cause rising upward from elementary particles to more and more complex matter.

vital body: the subtle world that contains the blueprints of form and function of organs that perform our fundamental vital functions.

vital energy: quantum movement of the vital body blueprints, the morphogenetic fields; called *prana* in India, *chi* in China, *ki* in Japan, or simply the life force in the West.

wave of possibility: many-faceted quantum object from which individual facets can be actualized through quantum measurement.

wave-particle duality: label created to mask the paradox that quantum objects seem to be both waves and particles.

weak objectivity: subjective experiences that are verified across a large number of subjects.

Further Reading

Aurobindo, S. (1996). *The Life Divine.* Pondicherry, India: Sri Aurobindo Ashram.
Briggs, J. (1990). *Fire in the Crucible.* Los Angeles, CA: Tarcher/Penguin.
Barrett, R. (1998). *Liberating the Corporate Soul.* Boston: Butterworth and Heinemann.
Capra, F. (1975). *The Tao of Physics.* Boulder, CO: Shambhala Press.
Chasse, Betsy, Mark Vicente, William Arntz. *What the Bleep Do We Know!?* Los Angeles: 20th Century Fox. 2005. DVD.
Chopra, D. (1990). *Quantum Healing.* New York: Bantam-Doubleday.
——— (2000). *Perfect Health.* New York: Three Rivers Press.
Clark, A. C. (1953). *Childhood's End.* New York: Ballantine Books.
Dossey, L. (1991). *Meaning and Medicine.* New York: Bantam.
Goswami, A. (1993). *The Self-Aware Universe: How Consciousness Creates the Material World.* New York: Tarcher/Putnam.
——— (1994). *Science Within Consciousness: A Monograph.* Petaluma, CA: Institute of Noetic Sciences.
——— (2000). *The Visionary Window: A Quantum Physicist's Guide to Enlightenment.* Wheaton, IL: Quest Books.
——— (2001). *Physics of the Soul.* Charlottsville, VA: Hampton Roads.
——— (2004). *The Quantum Doctor.* Charlottsville, VA: Hampton Roads.
——— (2008). *God Is Not Dead.* Charlottsville, VA: Hampton Roads.
——— (2008). *Creative Evolution.* Wheaton, IL: Theosophical Publishing House.

——— (2009). *The Quantum Activist*. Chehalis, WA: Intention Media, Inc. DVD.

——— (2011). *How Quantum Activism Can Save Civilization*. Charlottsville, VA: Hampton Roads.

——— (2014). *Quantum Creativity: Think Quantum, Be Creative*. New York: Hay House.

——— (2015). *Quantum Economics*. Faber, VA: Rainbow Ridge Books.

Herr, E. (2012). *Consciousness: Bridging the Gap Between Conventional Science and the Super Science of Quantum Mechanics*. Faber, VA: Rainbow Ridge Books.

McTaggert, L. (2007). *The Intention Experiments*. New York: Free Press.

Penrose, R. (1991). *The Emperor's New Mind*. New York: Penguin.

Pert, C. (1997). *Molecules of Emotion*. New York: Scribner.

Radin, D. (2009). *The Noetic Universe*. London: Transworld Publishers.

Searle, J. (1994). *The Rediscovery of the Mind*. Cambridge, MA: MIT Press.

Sheldrake, R. (1981). *A New Science of Life*. Los Angeles, CA: Tarcher.

Standish, L, J., Kozak, L., Clark Johnson, L., and Richards, T. (2004). "Electroencephalographic evidence of correlated event-related signals between the brains of spatially and sensory isolated human subjects." *The Journal of Alternative and Complementary Medicine*, vol. 10, 307–314.

Teilhard de Chardin, P. (1961). *The Phenomenon of Man*. New York: Harper & Row.

About the Author

Amit Goswami, PhD, is a theoretical nuclear physicist. He received his PhD in physics from Calcutta University in 1964 and has been a member of the University of Oregon's Institute of Theoretical Science since 1968. He is best known for his appearance as one of the interviewed scientists featured in the 2004 film *What the Bleep Do We Know!?*

Hampton Roads Publishing Company
. . . for the evolving human spirit

Hampton Roads Publishing Company
publishes books on a variety of subjects,
including spirituality, health, and other related topics.

For a copy of our latest trade catalog, call (978) 465-0504 or visit our distributor's website at *www.redwheelweiser.com*. You can also sign up for our newsletter and special offers by going to *www.redwheelweiser.com/newsletter/*.